₹20 —
633672

Lecture Notes in Economics and Mathematical Systems 614

Klaus Schredelseker · Florian Hauser

Complexity and Artificial Markets

Springer

Dr. Klaus Schredelseker
University of Innsbruck
Institute for Banking and Finance
Universitätsstr. 15
6020 Innsbruck
Austria
klaus.schredelseker@uibk.ac.at

Dr. Florian Hauser
University of Innsbruck
Institute for Banking and Finance
Universitätsstr. 15
6020 Innsbruck
Austria
florian.hauser@ uibk.ac.at

ISBN 978-3-540-70553-6 e-ISBN 978-3-540-70556-7

DOI 10.1007/978-3-540-70556-7

Lecture Notes in Economics and Mathematical Systems ISSN 0075-8442

Library of Congress Control Number: 2008930214

Production: le-tex Jelonek, Schmidt & Vöckler GbR, Leipzig
Cover design: WMX Design GmbH, Heidelberg

Printed on acid-free paper

9 8 7 6 5 4 3 2 1

springer.com

Preface

In 2000, when Levy, Levy, and Solomon published their book *Microscopic Simulation of Financial Markets*, Harry Markowitz noted in the blurb that numerical simulations point "us towards the future of financial economics. If we restrict ourselves to models which can be solved analytically, we will be modeling for our mutual entertainment, not to maximize explanatory or predictive power." At that time most economists were quite sceptical about the new techniques and thus a statement like this was encouraging for the Artificial Economics community. Since 2000, things have changed tremendously. Agent-based modeling, computer simulations, and artificial economics have become broadly accepted tools in social sciences by now. For a large number of problems they are the only reliable techniques to arrive at nontrivial results.

Neoclassical economics is usually split up into a micro and a macro analysis, the first dealing with the individual decision-maker (consumer, firm, investor etc.), and the second with economic aggregates such as aggregate demand and aggregate supply (labor, consumption, capital, etc.). The link, if there is any, between both levels is the representative agent, that is the assumption that either all agents are of the same type or that they act in such a way that the sum of their choices is mathematically equivalent to the decisions of identical, prototypical individuals. In such a world neither the problem of imperfect rationality nor the problem of disparate and diverse information can be addressed; the latter is not even the case if you allow for only two disparate levels of information, let us say informed and non-informed individuals.

What happens in the real world is an outcome of the interaction of numerous individuals, each of whom may have different preferences, different information levels and different attitudes. A system with a set of autonomous decision-makers (agents) who individually assess their situation and exhibit repetitive interactions based upon their idiosyncratic rules is called a multi-agent-model; it can give us valuable insights into the nature of the real system it attempts to emulate. If, as often has been formulated, a market is an open complex adaptive system with endogenous evolution, the only chance to get a deeper understanding of how it works will be to look at its dynamics, driven by individual decision making. If we do so, we possibly will capture emergent behavioral patterns which are the result of interaction and which are decoupled from the behavior of the individuals: the whole is more than its parts.

The 2008 meeting of researchers in Artificial Economics takes place in Innsbruck (Austria). The most distinguished scholar of our school was Eugen Ritter von Böhm-Bawerk, one of the protagonists of the so-called Austrian School of Economics. Agent-based modeling and complexity economics draw a lot of inspiration from and give a lot of inspiration to Austrian Economics. Central to this school of thought is the uncompromising use of methodological individualism and subjectivism: whatever happens in society has to be explained by going back to the actions of individuals; these individuals need not be perfectly rational, but they are assumed to exhibit at least meaningful and selfish decisions (as opposed to irrational agents, noise traders etc.). Austrian Economics always deals with 'human action' (Mises): It must be possible to explain why people do what they do. A theory, e.g. efficient markets theory, where in the end nobody has an incentive to do anything, is a sterile intellectual gimmick. In a perfect equilibrium as a result of competition, nobody competes; such theories are rather theories of human non-action than of human action. We are happy that some of the papers presented in the workshop will fit very well in the Böhm-Bawerk-tradition of Austrian Economics.

At the Innsbruck University School of Management agent-based modeling has quite a long tradition. The first paper of an artificial stock market with heterogeneously informed agents was published in 1997: it tried to resolve the famous information-paradox (Grossman/Stiglitz) without referring to market imperfections or to irrational decision making (as noise-traders) and showed that in a stock market you may be better off if you have less information than others (a typical result emerging from an agent-based approach). A lot of further work has been done in this field, partly using computer simulations and partly adopting an experimental approach. With respect to the objective of learning more about the dynamics of complex systems such as a market, both approaches have their merits, but also their shortcomings. Both stem from the dominant role of heterogeneous agents making individual decisions. If we want to gain reliable knowledge of how real human beings view their decision problems, which factors they take into account, how they deal with information overloads and other items, experimental economics with real people will be the more appropriate approach. If, however, we try to understand the underlying properties of a complex system, computer simulations will do better: with artificial agents we get economically 'pure' results which are not blurred by the bounded rationality of real agents. In both cases, however, macro phenomena grow on the sound ground of methodological individualism with autonomous agents; that is what counts.

All papers in this book have been selected in a double-blind reviewing process. They cover various topics within the area indicated in the title "Complexity and Artificial Markets".

The papers in Part I use agent-based simulations to deal with market mechanisms. The main concern of the *LiCalzi/Pellizzari*-paper is the market microstructure: how does resampling affect allocational efficiency in different market protocols? *Giulioni/Bucciarelli* observe the Pescara wholesale fish market with respect to its price dynamics. *Milone* studies the consequences of pre-trade quote disclosure on the market performance in different scenarios.

Part II is devoted to evolution and is decision making. *Anufriev and Hommes* show in an experimental study how different forecasting strategies perform in an evolutionary switching mechanism. *Raberto, Teglio, and Cincotti* focus on households' beliefs formation and financial preferences, based on concepts from prospect theory. *Fernández-de-Córdoba and Navas* present an evolutionary model and show under which conditions a Walrasian equilibrium is likely to emerge in an economy.

Garabedian presents an agent-based consumption model that is applied to the purchase decision for ethical goods.

Part III deals with information economics in a broad understanding. *Hule and Lawrenz* investigate the impact of information quality and the intensity of interaction on some stylized facts in financial markets. *Hofstede, Jonker, and Verwaart* create an agent-based model emphasizing the micro-dynamics of trust in a long-term trade relationship. Combining experimental economics and agent-based computational models *López-Paredes, Posada, Hernández, and Pajares* explain individual behavior of agents in a signaling game.

In Part IV, methodological issues prevail. *Livet, Phan, and Sanders* start from an ontological view and study the relationship between a given problem, experimental design, and modeling individual choice in different types of agent-based computational economics. *Van-der-Hoog, Deissenberg, and Dawid* present some new developments in the well-known agent-based model of the European economy called EURACE. *Grevers and Veen* compare the two main methodological approaches in social sciences, the systems approach and the individual-based approach, with special emphasis on agent-based computational economics.

It is almost a tradition of the Artificial Economics meetings to bring together people from computer science, natural sciences, philosophy, cognitive sciences, economics and finance, and other areas. The two invited speakers give evidence of this basically interdisciplinary approach. *Peter Henning*, coming from theoretical quantum physics, visited the world of financial markets at the Deutsche Börse AG, switched to computer science and, for the time being, teaches informatics, economics, e-learning and related fields. He, too, has a strong relationship to Tyrol as he supported for years the 'Bozner Treffen', an annual meeting of scientists coming from various disciplines. *Peter's* paper deals with different types of evolutionary processes: under which conditions can evolution serve as a bridge between biology and economics? *Alan Kirman* comes from neoclassical economics, but studying the link between micro and macro behavior he was a pioneer in agent-based computational economics; at an early stage he understood that economic activity is better viewed as the product of a complex self-organizing system than of corresponding to the behavior of an individual maximizer; with Innsbruck he is familiar as one of the speakers in the famous 'Böhm-Bawerk-lecture' given annually by some of the most distinguished economists from all over the world. *Alan* teaches at the Université de la Méditerranée near Marseille. His paper deals with rationality and organization in artificial markets.

Innsbruck,
May 2008

Klaus Schredelseker
Florian Hauser

Acknowledgements

We would like to thank all the members of the Scientific Committee who refereed the papers, gave most valuable comments to both editors and authors, and made it possible to publish this volume in time:

- **Frédéric Amblard**, Université Toulouse, France
- **Bruno Beaufils**, Université des Sciences et Technologies de Lille, France
- **Olivier Brandouy**, Université des Sciences et Technologies de Lille, France
- **Charlotte Bruun**, Aalborg University, Denmark
- **Andrea Consiglio**, Universita' degli Studi di Palermo, Italy
- **Wander Jager**, University of Groningen, The Netherlands
- **Marco Janssen**, Arizona State University, United States of America
- **Philippe Lamarre**, Université de Nantes, France
- **Michele Marchesi**, Università Cagliari, Italy
- **Luigi Marengo**, Sant'Anna School of Advanced Studies, Italy
- **Philippe Mathieu**, Université de Lille 1, France
- **Nicolas Maudet**, Université Paris-Dauphine, France
- **Akira Namatame**, National Defense Academy, Japan
- **Paolo Pellizzari**, Università "Ca' Foscari" di Venezia, Italy
- **Denis Phan**, GEMAS CNRS & Université Paris IV Sorbonne, France
- **Juliette Rouchier**, GREQAM, France
- **Enrico Scalas**, Università del Piemonte Orientale, Italy
- **Elpida Tzafestas**, National Technical University of Athens, Greece
- **Murat Yildizoglu**, Université Montesquieu Bordeaux IV, France
- **Stefano Zambelli**, Trento University, Italy

We acknowledge financial support for the conference by the **Austrian Bundesministerium für Wissenschaft und Forschung** and by the **Vizerektor für Forschung at the University of Innsbruck**. Without these grants the realization of this conference would not have been possible. We also thank **Philip Herdina** for helping us with proofreading the papers.

Contents

Contributors

Mikhail Anufriev
CeNDEF, University of Amsterdam, Roetersstraat 11, NL-1018 WB, Amsterdam, the Netherlands, e-mail: M.Anufriev@uva.nl

Edgardo Bucciarelli
Department of Quantitative Methods and Economic Theory, University of Chieti-Pescara, Viale Pindaro 42, 65127 Pescara, Italy, e-mail: e.bucciarelli@unich.net

Silvano Cincotti
DIBE-CINEF, University of Genova, Via Opera Pia 11a, 16145 Genova, Italy, e-mail: cincotti@dibe.unige.it

Herbert Dawid
Bielefeld University, Dept. of Business Administration and Economics, Universitätsstrasse 25, D-33615 Bielefeld, Germany, e-mail: hdawid@wiwi.uni-bielefeld.de

Christophe Deissenberg
Université de la Méditerranée II and GREQAM, Château Lafarge, Route des Milles, 13290 Les Milles, France, e-mail: christophe.deissenberg@univmed.fr

Gonzalo Fernández-de-Córdoba
Departamento de Economía e Historia Económica, Universidad de Salamanca, Edificio F.E.S. Campus Miguel de Unamuno, E-37007 Salamanca, Spain, e-mail: gfdc@usal.es

Sabine Garabedian
GREDEG, University of Nice Sophia-Antipolis-CNRS, 250 av. Albert Einstein, 06650 Valbonne, France, e-mail: garabedi@gredeg.cnrs.fr

Gianfranco Giulioni
Department of Quantitative Methods and Economic Theory, University of Chieti-Pescara, Viale Pindaro 42, 65127 Pescara, Italy, e-mail: g.giulioni@unich.it

Wilbert Grevers
University of Groningen, P.O. Box 716, 9700 AS Groningen, The Netherlands, e-mail: w.a.j.grevers@rug.nl

Peter A. Henning
Institute for Computers in Education, Karlsruhe University of Applied Sciences, Moltkestrasse 30, 76133 Karlsruhe, Germany, e-mail: peter.henning@ice-karlsruhe.de

Cesáreo Hernández
Valladolid INSISOC, Valladolid University, Paseo del Cauce s/n 47011 Valladolid, Spain, e-mail: cesareo@eis.uva.es

Gert J. Hofstede
Wageningen University, Postbus 9109, 6700 HB Wageningen, The Netherlands, e-mail: gertjan.hofstede@wur.nl

Cars Hommes
CeNDEF, University of Amsterdam, Roetersstraat 11, NL-1018 WB, Amsterdam, the Netherlands, e-mail: C.H.Hommes@uva.nl

Sander van der Hoog
Université de la Méditerranée II and GREQAM, Château Lafarge, Route des Milles, 13290 Les Milles, France, e-mail: svdhoog@gmail.com

Richard Hule
Department of Economics, Innsbruck University, Universitätsstrasse 15, 6020 Innsbruck, Austria, e-mail: richard.hule@uibk.ac.at

Catholijn M. Jonker
Delft University of Technology, Mekelweg 4, 2628 CD Delft, The Netherlands, e-mail: C.M.Jonker@tudelft.nl

Alan Kirman
GREQAM, Université de la Méditerranée, 2 rue de la Charité, 13002 Marseille, France, e-mail: alan.kirman@univmed.fr

Jochen Lawrenz
Department of Banking & Finance, Innsbruck University, Universitätsstrasse 15, 6020 Innsbruck, Austria, e-mail: jochen.lawrenz@uibk.ac.at

Marco LiCalzi
Department of Applied Mathematics and SSE, Università Ca' Foscari di Venezia, Dorsoduro 3825/E 30123, Venice, Italy, e-mail: licalzi@unive.it

Pierre Livet
CEPERC, UMR 6059, CNRS & Université de Provence, France, e-mail: livet@up.univ-mrs.fr

Adolfo López-Paredes
Valladolid INSISOC, Valladolid University, Paseo del Cauce s/n 47011 Valladolid, Spain, e-mail: adolfo@insisoc.org

Lucia Milone
Department of Economics, University of Venice, Cannaregio 873, 30121 Venice, Italy, e-mail: lucia.milone@unive.it

Álvaro P. Navas
UAV Navigation SL. Av. Severo Ochoa no45, Alcobendas, Madrid, 28100, Spain, e-mail: anavas@uavnavigation.com

Javier Pajares
Valladolid INSISOC, Valladolid University, Paseo del Cauce s/n 47011 Valladolid, Spain, e-mail: pajares@eis.uva.es

Paolo Pellizzari
Department of Applied Mathematics and SSE, Università Ca' Foscari di Venezia, Dorsoduro 3825/E 30123, Venice, Italy, e-mail: paolop@unive.it

Denis Phan
GEMAS UMR 8598, CNRS & University Paris IV - Sorbonne, France, e-mail: dphan@msh-paris.fr

Marta Posada
Valladolid INSISOC, Valladolid University, Paseo del Cauce s/n 47011 Valladolid, Spain, e-mail: posada@eis.uva.es

Marco Raberto
DIBE-CINEF, University of Genova, Via Opera Pia 11a, 16145 Genova, Italy, e-mail: raberto@dibe.unige.it

Lena Sanders
Géographie-cités, UMR 8504, CNRS & Université Paris 1 & Université Paris 7, France, e-mail: lena.sanders@parisgeo.cnrs.fr

Andrea Teglio
DIBE-CINEF, University of Genova, Via Opera Pia 11a, 16145 Genova, Italy, e-mail: teglio@dibe.unige.it

Anne van der Veen
International Institute for Geo-Information Science and Earth Observation, P.O. Box 6, 7500 AA Enschede, The Netherlands, e-mail: veen@itc.nl

Tim Verwaart
LEI Wageningen UR, Postbus 29703, 2502 LS, Den Haag, the Netherlands, e-mail: tim.verwaart@wur.nl

Part I
Market Mechanisms

Chapter 1
Zero-Intelligence Trading Without Resampling

Marco LiCalzi and Paolo Pellizzari

Abstract This paper studies the consequences of removing the resampling assumption from the zero-intelligence trading model in Gode and Sunder (1993). We obtain three results. First, individual rationality is no longer sufficient to attain allocative efficiency in a continuous double auction; hence, the *rules of the market* matter. Second, the allocative efficiency of the continuous double auction is higher than for other sequential protocols both with or without resampling. Third, compared to zero intelligence, the effect of learning on allocative efficiency is sharply positive without resampling and mildly negative with resampling.

1.1 Introduction

In a recent paper, Mirowski (2007) argues that we are witnessing a "shift to a market-centered theory of computational economics" (p. 214). He attributes an important strand in this shift to the ramifications of Gode and Sunder (1993). This seminal paper is widely credited[1] with showing that the continuous double auction can attain allocative efficiency and convergence to the equilibrium price in the absence of trader intelligence. Such zero-intelligence (henceforth, ZI) is modeled by replacing human subjects with computerized agents that generate random quotes.

As Mirowski himself acknowledges, "there is still substantial dispute over the interpretation of their results" (p. 216); e.g., see Brewer et al. (2002). The boldest claim is that an appropriate market institution can override the cognitive limitations of individuals to achieve allocative efficiency and discover the equilibrium price. On the other side of the fence, the sharpest criticism is offered by Gjerstad and Shachat

[1] See Footnote 5 in Gjerstad and Shachat (2007).

M. LiCalzi and P. Pellizzari
Department of Applied Mathematics and SSE, Università Ca' Foscari di Venezia, Dorsoduro 3825/E 30123, Venice, Italy, e-mail: [licalzi,paolop]@unive.it

(2007). This paper provides a fresh and careful reading of Gode and Sunder (1993) that makes two points: first, convergence to the equilibrium price does not actually occur in Gode and Sunder (1993); second, the key condition for allocative efficiency is the *individual rationality* of the agents rather than the *market discipline* imposed by the continuous double auction.

Based on this, Gjerstad and Shachat (2007) conclude that "individual rationality is both necessary and sufficient to reach" allocative efficiency (p. 7). This argument is backed up by the claim that Gode and Sunder (1993) deal with a special case of the B-process for which Hurwicz et al. (1975) prove that in an economy without externalities a random but otherwise individually rational behavior converges to a Pareto optimal allocation.

In fact, this claim rests on a subtle but far from innocuous assumption made in Gode and Sunder (1993) that has gone largely unnoticed in the literature. We quote from Gode and Sunder (1993, p. 122): "There are several variations of the double auction. We made three choices to simplify our implementation of the double auction. Each bid, ask, and transaction was valid for a single unit. *A transaction canceled any unaccepted bids and offers.* Finally, when a bid and a ask crossed, the transaction price was equal to the earlier of the two." (Emphasis added.) We call the emphasized assumption *resampling* because under zero intelligence it forces all agents who have already uttered a quote to issue a new (random) one after each transaction.

This paper studies the consequences of removing the resampling assumption. We obtain three results. First, under zero intelligence, individual rationality without resampling is not sufficient to attain allocative efficiency in a continuous double auction; hence, the *rules of the market* matter. On the other hand, with or without resampling, the allocative efficiency of the continuous double auction is higher than for the other sequential protocols we consider; hence, this market protocol is still the most effective among those. Third, when zero intelligence is replaced by a simple variant of the algorithm mimicking learning-based human behavior proposed in Gjerstad and Dickhaut (1998), we find that the effect on allocative efficiency is sharply positive without resampling but tends to be mildly negative with resampling.

1.2 The Model

We use a setup inspired to Gode and Sunder (1993). There is an economy with a large number ($n = 5000$) of traders, who can exchange single units of a generic good. Each agent is initialized to be a seller or a buyer with equal probability. Each seller i is endowed with one unit of the good for which he has a private cost c_i that is independently drawn from the uniform distribution on $[0,1]$. Each buyer j holds no units and has a private valuation v_j for one unit of the good that is independently drawn from the uniform distribution on $[0,1]$. By individual rationality, each seller i is willing to sell his unit at a price $p \geq c_i$ and each buyer j is willing to buy at most one unit at a price $p \leq v_j$.

Gode and Sunder (1993) make the three simplifying assumptions cited above. We maintain the first one and restrict all agents to trade at most one unit. The third assumption selects the continuous double auction as the market protocol that regulates the interactions between traders. We expand on this and compare the allocative efficiency of four different sequential protocols, including of course the continuous double auction. These four protocols are: the continuous double auction, a nondiscretionary dealership, a hybrid of these two, and the trading pit. The first three are described in detail in LiCalzi and Pellizzari (2006, 2007a).

Briefly, in the continuous double auction (henceforth C) traders sequentially place quotes on the selling and buying books. Orders are immediately executed at the outstanding price if they are marketable; otherwise, they are recorded on the books with the usual price-time priority and remain valid until the end of the trading session. In the trading pit (henceforth T), traders are randomly matched in pairs: each agent in a pair utters a quote and, if compatible, they transact at a price equal to the average of their quotes; this transaction price is made known to the market, but its participants have no access to the offers exchanged within a pair.

In the dealership (henceforth, D) there is a specialist who posts bid and ask quotes valid only for a unit transaction. Agents arrive sequentially and can trade only at the dealer's quotes. Right after a transaction is completed, both dealer's quotes increase (or decrease) by a fixed amount k when the agent completes a purchase (or a sale); hence, the bid-ask spread Δ remains constant. Clearly, completing a trade between a buyer and a seller by going through the dealer is costly: for instance, if trader i sells one unit to the dealer that is immediately after resold to buyer j, the dealer pockets a value of $\Delta - k$. In this respect, the presence of the dealer negatively affects allocative efficiency. On the other hand, because the dealer guarantees a fixed bid-ask spread, it has a stabilizing effect on price dispersion that is usually beneficial.

For a large range of values, the force of these two effects vary in a predictable manner. Hence, the instantiation of k and Δ is influential but not crucial: we assume $k = 0.005$ and $\Delta = 0.05$ throughout the paper. The choice of the initial dealer's quotes, instead, is more delicate: when these happen to be far away from the equilibrium price, the effect on allocative efficiency may be relevant because the first few trades tend to occur on the same side of the market (until the dealer's quotes get closer to the equilibrium price). Except for a final comment in Sect. 1.3.3, we mute this issue and assume that the initial quotes exactly straddle the (theoretical) equilibrium price. Finally, the hybrid market (henceforth, H) combines the continuous double auction with the dealership: agents have access to the dealer's quotes as well as to the offers from the public recorded in the book. The initialization for H is the same used for D; that is, $k = 0.005$, $\Delta = 0.05$ and the initial dealer's quotes straddle the equilibrium price.

Each of these four protocols is organized over a single trading session, where all agents participate. Their order of arrival is randomly selected. Whenever a transaction takes place between two agents, their own orders are removed from the market and the agents become inactive. The difference between assuming resampling or not is the following. Under no resampling, each agent gets only one chance to act: he can trade up to one unit (if he finds a suitable quote) or, limitedly to C and H, utter his own quote (that remains valid until the end). The market closes after all agents have had their chance to act. Under resampling as postulated in Gode and Sunder (1993), until an agent completes a trade and becomes inactive, the refresh following a trade may give him a new chance to act. Therefore, the number of chances for actions is much greater under resampling, and this tends to increase allocative efficiency. To minimize this bias, we assume that under resampling the market closes when, following a refresh, all the active agents have issued a quote and no transaction has occurred.

Two more differences separate the book-based (we call them "literate") protocols C and H from the "illiterate"[2] protocols D and T. First, the book in C and H offers to the current agent an option to store his quote, extending his opportunity to trade in the future; on the contrary, D and T limit his options to immediate trade or no trade at all. Second, the book makes quotes from past traders available to the current agent, presenting him with a larger set of potential counterparts for his trade; on the other hand, for illiterate protocols the only available counterpart is the dealer in D and a single partner in T. In other words, a literate protocol expands the opportunities for trades as well as the pool of potential counterparts. These differences are not a crucial issue under resampling, because a trader returning to the market faces a new opportunity to trade, usually at different conditions. However, as we discuss below, they have a substantial effect when resampling is not allowed.

We use two different behavioral assumptions in our tests. Under zero intelligence, when an agent i must issue a quote, he picks a random number from the uniform distribution on $[0, v_i]$ if he is a buyer and from the uniform distribution on $[c_i, 1]$ if he is a seller. This behavior corresponds to zero intelligence under individual rationality and is called ZI-C in Gode and Sunder (1993). The second behavioral assumption is a simplified version[3] of the learning model introduced in Gjerstad and Dickhaut (1998), where each trader transforms the empirical acceptance frequencies to generate beliefs and then issues the quote that maximizes his expected surplus with respect to these beliefs. This approach is in general quite sensitive to fine details in its initialization and implementation. However, it can be calibrated to effectively mimic the basic features of human behavior in experimental trading markets. See Gjerstad (2007) for more details and an improved version of the original (1998) model.

Our implementation is the following. We discretize the unit interval $[0, 1]$ for prices by assuming a "tick" equal to $1/200 = 0.005$. Let $H_t(x) = \#(p \leq x)$ denote the empirical cumulative frequency of past transaction prices at time t. Each buyer i starts up with a uniform "prior" described by the cumulative distribution $F_i(x) = \min\{(x - nbv_i)^+, 1\}$ on the ticked prices contained in the interval $[bv_i, v_i]$, where $b = 0.8$. (For a seller i, we assume by symmetry a uniform distribution over the interval $[c_i, (1-b) + bc_i]$.) This initial distribution is associated to a coefficient a_i that defines the stubbornness of i's initial beliefs; we assume that a_i is an integer drawn (once for each agent) from the uniform distribution on $\{1, 2, \ldots, 100\}$. When

[2] This terminology is non-standard, but less convoluted than a plain "non-book-based."

[3] The most notable difference is that we do not assume bounded recall of past transactions.

a buyer i is called up for trading at time t, he combines his "prior" with the empirical distribution $H_t(p)$ and derives a "posterior" cumulative distribution $P(p \leq x)$ that is proportional to $a_i F_i(x) + H_t(x)$. Then buyer i issues a bid b that maximizes his expected utility $(v - b) \cdot P(p \leq b)$. (Sellers' behavior is analogous.)

1.3 Results

We are interested in the allocative efficiency of different market protocols under zero intelligence. As usual, allocative efficiency is defined as the ratio between the realized gains from the trade and the maximum feasible gains from trade. This measure is adimensional, facilitating comparisons. We compare the allocative efficiency of the four protocols described above under both zero intelligence and our version of the learning model proposed in Gjerstad and Dickhaut (1998). Since we view the role of the dealer as a mere feature of the protocol, his final gains/losses are not included in the computation of the allocative efficiency.

1.3.1 Test 1: Does Resampling Matter?

Assume zero-intelligence trading. The left-hand side of Fig. 1.1 shows as datapoints the allocative efficiency of the continuous double auction *with resampling* for 100 different runs. The right-hand side shows the same information for the continuous double auction *without resampling*. The y-axes use the same scale, so that it is possible to directly compare the results under the two assumptions by visual inspection: the higher the level, the higher the allocative efficiency.

The average allocative efficiency is 0.96 with resampling and 0.52 without resampling. Visual inspection strongly suggests that the distribution of the allocative efficiency with resampling stochastically dominates the distribution without

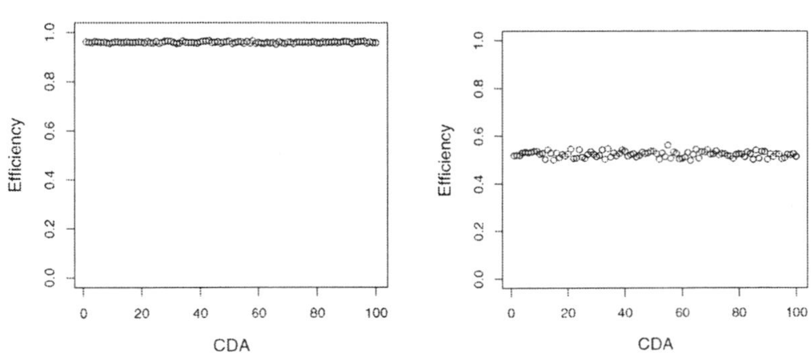

Fig. 1.1 Allocative efficiency for C with (*left*) and without resampling (*right*)

resampling. More modestly, we claim that under resampling the expected value of allocative efficiency is higher. This is supported for any practical level of confidence by the directional version of the Wilcoxon signed-rank test. (Here and in the following, by a *practical* level of confidence we mean a p-value lower than 10^{-5}.) Limited to our experiment, therefore, we conclude that *ceteris paribus* resampling yields a higher allocative efficiency than no resampling. In short, resampling truly matters a lot.

1.3.2 Test 2: Which Protocol Performs Better Under Zero Intelligence?

Our second test extends the first one by looking at the effects of resampling under zero intelligence for other sequential protocols. Each protocol is identified by its initials on the x-axis and by a different color: the continuous double auction (C) is in black; the nondiscretionary dealership (D) is in red; the hybridization (H) of the continuous double auction with a dealership is in green; and the trading pit (T) is in blue. The left-hand side of Fig. 1.2 reports for each protocol the allocative efficiency *with resampling* for 100 different runs, as well as the sample average at the bottom of each column. The right-hand side shows the same information for the continuous double auction without resampling. Again, the y-axes use the same scale so direct comparison is possible.

We make two claims. The first one is that, for each protocol, allocative efficiency with resampling is significantly much higher than without resampling. This confirms and reinforces our earlier claim that the assumption of resampling matters a lot. The data in black concerning the continuous double auction (C) are reproduced from Fig. 1.1 and need no further commentary. The data in red regarding the dealership (D) report a sample average of 0.91 with resampling against 0.33 with no resampling. Analogously, the data in green regarding the hybrid protocol (H) give a sample average of 0.94 with resampling against 0.42 with no resampling. Finally,

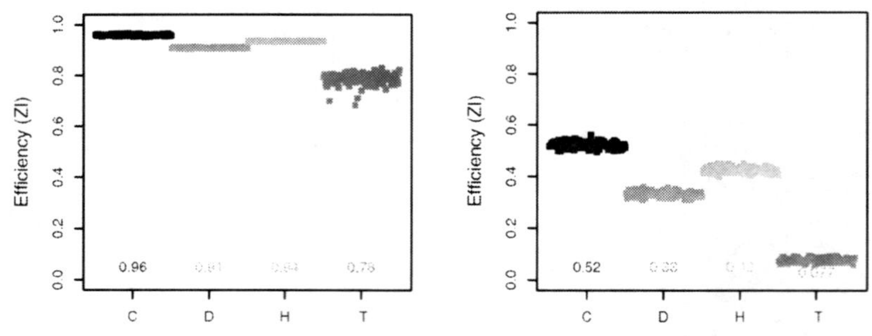

Fig. 1.2 Allocative efficiency with (*left*) and without resampling (*right*)

the sample averages for the trading pit (T) is 0.78 with resampling and 0.077 with no resampling. For each protocol, the directional version of the Wilcoxon signed-rank test supports the significance of the difference with and without resampling for any practical level of confidence.

We conclude that the introduction of the resampling assumption has a dramatic positive effect on allocative efficiency under zero intelligence. Hence, this assumption introduces an important bias that undermines Gode and Sunder's (1993) claim that "the primary cause of the high allocative efficiency of double auctions is the market discipline imposed on traders" (p. 134), unless such market discipline is not taken to include resampling as well.

Two minor observations are worth making. First, regardless of the resampling assumptions, allocative efficiency is higher for literate protocols. The reason is that they give each trader access to a larger pool of counterparts. Second, the differences in the allocative efficiency of the trading pit are exaggerated by the minor modeling assumption that traders are matched in pairs. This implies that several pairs end up being formed by traders on the same side of the market who are bound not to trade. Therefore, we have also tested the alternative assumption that buyers and sellers are matched in pairs, making sure that each pair is formed by agents on the opposite side of the market. In this second case, the sample average of the allocative efficiency without resampling is 0.15. No qualitative conclusion is affected, although it is obvious that the trading pit works much better if traders can be screened in buyers versus sellers before being matched. (For each of the other protocols, the adirectional version of the Wilcoxon signed-rank test supports at any practical level of confidence the claim that it is makes no difference for allocative efficiency to have buyers and sellers arrive in random order or alternately.)

Our second claim is that the allocative efficiency of the continuous double auction with or without resampling is higher than for other sequential protocols; hence, this market protocol remains more effective under zero intelligence. This is easily detectable by visual inspection of the two tables in Fig. 1.2. The directional version of the Wilcoxon signed-rank test supports the claim that C yields a higher expected allocative efficiency than H (the highest competitor) for any practical level of confidence, both in the case of resampling (left) and no resampling (right). This confirms Gode and Sunder's (1993) intuition that the continuous double auction provides an important and natural benchmark for allocative efficiency under zero intelligence. The next test inquires whether this remains true under more realistic assumptions about agents' behavior.

1.3.3 Test 3: Does Learning Make a Difference?

Our third test extends the previous one by looking at the allocative efficiency of protocols under the alternative assumption that traders learn and optimize according to a slightly simplified version of the model in Gjerstad and Dickhaut (1998). We consider first the case without resampling, and then the case with resampling.

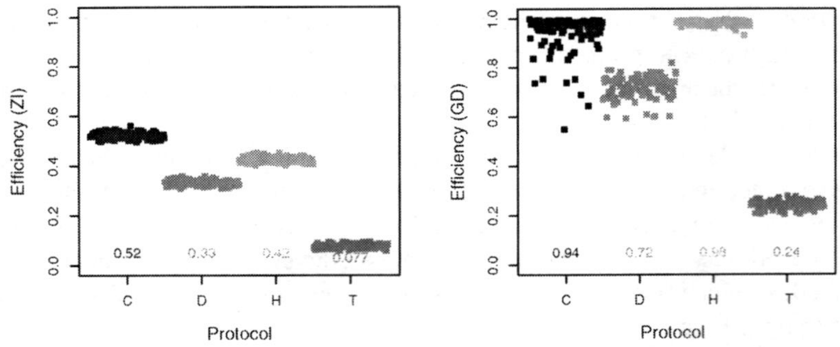

Fig. 1.3 Allocative efficiency under heuristic learning without resampling

The right-hand side of Fig. 1.3 shows for each protocol the allocative efficiency without resampling for 100 different runs as well as the sample average, under the assumption that all agents base their trading on our simple model of learning and optimization. The left-hand side shows the same information (copied from Fig. 1.2) under zero-intelligence trading for ease of comparison. The usual coding applies for initials and color.

Direct inspection shows that, without resampling, learning greatly improves the expected value of the allocative efficiency for each protocol. The sample average shoots from 0.52 to 0.94 for C, from 0.33 to 0.72 for D, from 0.42 to 0.98 for H, and from 0.077 to 0.24 for T. (The directional Wilcoxon test supports this hypothesis for any practical level of confidence.) This effect is easily explained. Under zero intelligence, resampling implies a complete refresh after each trade, in the sense that (conditional on the set of active traders) the probability distribution of the next quote is independent of the past. This makes the prices of the first few transactions almost irrelevant for predicting future behavior. On the other hand, under learning, the prices of past transactions affect the beliefs and hence the actions of future traders. The initial transaction prices feed future beliefs, amplifying the effect.

Even tough the lack of resampling forbids agents from revising their past quotes, the learning process substitutes for this because incoming agents use past history when formulating their quotes. Therefore, under no resampling, learning makes a huge difference for the allocative efficiency of a market protocol. To the extent that learning is a behavioral assumption while no resampling is an institutional feature, this strongly suggests that we cannot apply the claim that "the primary cause of the high allocative efficiency of double auctions is the market discipline imposed on traders" (Gode and Sunder, 1993, p. 134) to situations in which resampling does not hold.

A comparison of the left- and right-hand sides of Fig. 1.3 shows two more effects. First, regardless of the behavioral assumptions, allocative efficiency is as usual higher for literate protocols. Second, learning-based behavior tends to increase the dispersion of allocative efficiency. The sample standard deviation goes from 0.013 to 0.081 for C, from 0.011 to 0.046 for D, from 0.011 to 0.010 for H, and from 0.008

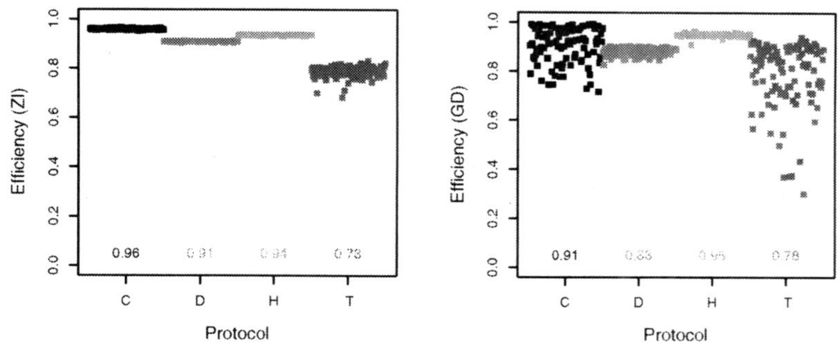

Fig. 1.4 Allocative efficiency under heuristic learning with resampling

to 0.016 for T. There is a sharp increase for three protocols and a mild decrease for one. This is another manifestation of the path-dependency implicit in our learning process: when a few initial trades off the equilibrium price point beliefs in the wrong direction, behavior may cluster around the wrong price and reduce allocative efficiency. This tends to increase the variability in performance, although the overall effect remains clearly favorable.

We now move to consider resampling. Figure 1.4 reports the same information as Fig. 1.3 under the assumption of resampling. The left-hand side assumes zero intelligence; the right-hand side is based on our learning-based behavioral assumption. The usual coding applies. Once again, regardless of the behavioral assumptions, allocative efficiency is higher for literate protocols.

Under resampling, learning degrades the expected allocative efficiency for three protocols. The sample average falls from 0.96 to 0.91 for C, from 0.91 to 0.88 for D, and from 0.7845 to 0.7838 for T, while it increases from 0.94 to 0.95 for H. (For C, D, and T, the directional Wilcoxon test supports the claim for any practical level of confidence; for H, the confidence level is 0.0007.) The sample standard deviation goes from 0.003 to 0.074 for C, from 0.002 to 0.017 for D, from 0.0019 to 0.006 for H, and from 0.024 to 0.139 for T.

Compared to the sharp improvement it carries without resampling, learning tends to have an opposite effect and bring about a reduction on allocative efficiency under resampling. Given the extremely high values of allocative efficiency under zero intelligence, this is not surprising. There is very little room to improve on allocative efficiency, so the rare occasions when path-dependent beliefs fixate on the wrong price end up reducing the allocative efficiency and increasing its dispersion. This is however not true for the hybrid protocol, because it can exploit the traders' book to reduce the amount of allocative efficiency lost to the dealer as well as the presence of the dealer herself to reduce the chance of transaction prices fixating on the wrong price.

The effects of path-dependent learning can also be apprised by comparing the time series of the transaction prices. The left-hand (respectively, right-hand) side of Fig. 1.5 shows two representative series for the continuous double auction with

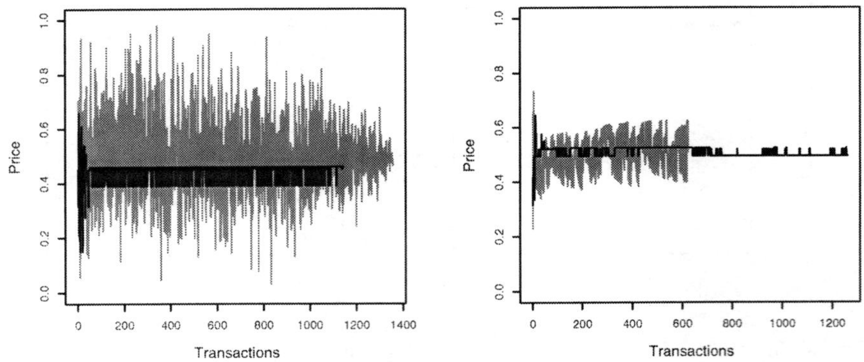

Fig. 1.5 Series of transaction prices for C with (*left*) and without resampling (*right*)

(without) resampling: zero intelligence is in red, learning-based behavior is in black. The x-axis reports the numbers of transactions: the longer the series, the higher the volume.

Zero intelligence exhibits comparatively wilder oscillations that eventually tend to fade out, but are centered around the correct equilibrium price. Moreover, as obvious, the volume generated with resampling is much higher than without resampling. Finally, it is apparent that lack of resampling sharply reduces the overall dispersion of transaction prices. A careful look at the right-end tail of the series on the left shows that the dispersion with resampling becomes comparable to that one without resampling precisely when trading in the latter approaches its end.

Learning-based behavior generates much tighter series around some price. Allocative efficiency is hurt in those relatively unfrequent cases where this is different from (on the left of Fig. 1.5, lower than) the equilibrium price. Independently of the assumptions about resampling, whenever beliefs fixate around the "wrong" price, the volume of transactions goes down and this hurts allocative efficiency. But there are two crucial differences. First, without resampling, traders cannot re-enter the market after having issued a quote: as wrong beliefs affect less people, the effect is reduced. Second, and limitedly to the two illiterate protocols (D and T), volume without resampling is heavily impaired by the combination of potentially wrong beliefs with uniqueness of the opportunity to trade and of the potential counterpart. As a result, given learning based-behavior, giving up resampling has a moderate positive effect for allocative efficiency in the two literate protocols C and H and a large negative effect for the illiterate protocols D and T.

A related issue is the robustness of our learning-based model for protocols involving a dealer in extreme situations, such as when the initial dealer's quote are very far from the equilibrium price. For instance, Table 1.1 summarizes the sample averages of allocative efficiency for D and H with and without resampling as well as under zero intelligence or learning-based behavior under two different initializations. On the left we report the case where the initial dealer's quotes straddle the equilibrium price $p = 0.5$, as assumed throughout this paper. On the right, we report the case

Table 1.1 The impact of the dealer's initial quotes on allocative efficiency

	$p = 0.5$				$p = 0.03$			
	Zero intelligence		Learning		Zero intelligence		Learning	
	D	H	D	H	D	H	D	H
Without resampling	0.33	0.42	0.72	0.98	0.37	0.49	0.10	0.23
With resampling	0.91	0.94	0.88	0.95	0.94	0.97	0.13	0.26

where the initial dealer's quotes straddle $p = 0.03$. It is clear that the quantitative effects of extreme initializations for D and H may be substantial.

To sum it up, compared to zero-intelligence, learning-based behavior brings about two effects on the allocative efficiency of a protocol. The negative one is the occasional clustering of path-dependent beliefs around the wrong price. This reduces allocative efficiency. The positive effect is that beliefs fixating around the equilibrium price help future traders to avoid wrong quotes. This improves allocative efficiency. Without resampling, the positive effect swamps the negative one for all protocols. Under resampling, the negative effect tends to prevail. (An additional effect related to the literacy of a protocol emerges when comparing allocative efficiency under learning-based behavior with and without resampling.) While this is not the place for sweeping generalizations, it seems legitimate to conjecture that the performance of a protocol under zero intelligence *with* resampling is not a good proxy for its performance under learning-based human behavior *without* resampling.

1.4 Conclusions

Our first result is that, under zero intelligence, individual rationality without resampling is not sufficient to attain allocative efficiency in the continuous double auction. Stronger than that, it is not sufficient in none of the four market protocols we study: none of the simulations reported on the left achieves more than 60% of the maximum allocative efficiency. This establishes that the *rules of the market* matter: when resampling is ruled out, zero intelligence rules out allocative efficiency. The "second" assumption in Gode and Sunder (1993) is not a mere simplification, but an important restriction.

Moreover, as shown in Sect. 1.3.2, assuming resampling on top of zero intelligence leads to a sharp increase in allocative efficiency for any of the four different protocols we tested. Therefore, the assumption of no resampling introduces a substantial bias towards achieving allocative efficiency. Gode and Sunder's (1993) claim that "the primary cause of the high allocative efficiency of double auctions is the market discipline imposed on traders" (p. 134) does not hold when their unconspicuous assumption of resampling is dropped.

On the other hand, we also find that the allocative efficiency of the continuous double auction under zero intelligence is never lower than for the other three market protocols. This validates Gode and Sunder's (1993) intuition about the effectiveness of the continuous double auction. In some cases, there may be other market protocols that exhibit similar results in this respect. This suggests the importance of introducing additional performance measures to rank market protocols that exhibit a similar degree of allocative efficiency, as discussed in LiCalzi and Pellizzari (2007b).

Our final result concerns the performance of the four market protocols when zero intelligence is replaced by a simple version of the learning-based model of human behavior proposed in Gjerstad and Dickhaut (1998). We find that learning-based behavior has two opposite effects on allocative efficiency. When path-dependent beliefs end up clustering around the wrong price, allocative efficiency is reduced. When instead they fixate around the equilibrium price allocative efficiency improves. The sign for the combination of these two effects is *a priori* ambiguous. We find that it is sharply positive without resampling and (usually) mildly negative with resampling. Since human behavior is very likely to have more in common with learning than with zero intelligence, this implies that the ability of a protocol to steer human traders towards allocative efficiency is not independent of the assumptions about resampling.

References

P.J. Brewer, M. Huang, B. Nelson, and C.R. Plott. On the behavioral foundations of the law of supply and demand: Human convergence and robot randomness. *Experimental Economics*, 5: 179–208, 2002.

S. Gjerstad. The competitive market paradox. *Journal of Economic Dynamics and Control*, 31: 1753–1780, 2007.

S. Gjerstad and J. Dickhaut. Price formation in double auctions. *Games and Economic Behavior*, 22:1–29, 1998.

S. Gjerstad and J.M. Shachat. Individual rationality and market efficiency. wp 1204, August. IRBEMS, Krannert School, Purdue University, 2007.

D.K. Gode and S. Sunder. Allocative efficiency of markets with zero intelligence traders: Market as a partial substitute for individual rationality. *Journal of Political Economy*, 101:119–137, 1993.

L. Hurwicz, R. Radner, and S. Reiter. A stochastic decentralized resource allocation process: Part I. *Econometrica*, 43:187–222, 1975.

M. LiCalzi and P. Pellizzari. The allocative effectiveness of market protocols under intelligent trading. In C. Bruun, editor, *Advances in Artificial Economics*, pages 17–29. Springer, Berlin, 2006.

M. LiCalzi and P. Pellizzari. Simple market protocols for efficient risk sharing. *Journal of Economic Dynamics and Control*, 31:3568–3590, 2007a.

M. LiCalzi and P. Pellizzari. Which market protocols facilitate fair trading? In A. Consiglio, editor, *Artificial Markets Modeling*, pages 81–97. Springer, Berlin, 2007b.

P. Mirowski. Markets come to bits: Evolution, computation and markomata in economic science. *Journal of Economic Behavior and Organization*, 63:209–242, 2007.

Chapter 2
Understanding the Price Dynamics of a Real Market Using Simulations: The Dutch Auction of the Pescara Wholesale Fish Market

Gianfranco Giulioni and Edgardo Bucciarelli

Abstract This paper aims to contribute towards the literature on artificial market modeling of real estate markets. Having the possibility to observe the Pescara wholesale fish market, we build an agent based simulation model to find out which buyers' behavior is able to replicate its price dynamics. The present work differentiates itself from other works on perishable products because the market under investigation has a centralized structure organized as a Dutch auction instead of being based on bilateral trading. From this point of view the paper also contributes to understand the anomalies of price dynamics highlighted by the empirical papers on sequential auctions for art, wine and jewelry.

2.1 Introduction

Real economic contexts often deliver different results from those foreseen by theoretical models. It is probably because agents, whose interaction determine the final results, move in an ever changing environment that prevents them from performing in a way which maximizes their objectives as described in the standard theoretical analysis. In this context it is plausible that agents' behavior evolves adaptively or they imitate their neighbors looking for utility or profit improvements.

Artificial market modeling is one (and perhaps the only) way to investigate these situations. A large number of studies have concentrated their attention on financial markets. They mostly attempt to replicate the odd price dynamics observed in real data; a celebrated example is the Santa Fe Artificial Stock Market (Arthur et al., 1997). Real markets receive little attention from this point of view mainly because of lack of data. A number of works study perishable products wholesale markets

G. Giulioni and E. Bucciarelli
Department of Quantitative Methods and Economic Theory, University of Chieti-Pescara, Viale Pindaro 42, 65127 Pescara, Italy, e-mail: g.giulioni@unich.it, e.bucciarelli@unich.net

(Kirman and Vignes, 1991; Graddy, 1995; Kirman et al., 2005). The cited papers contain empirical investigations that basically detect the presence of imperfect competition in markets where exchanges take place after a bilateral bargaining process. Artificial modeling of real markets makes up a few papers that concentrate on agents' learning to trade in a situation of imperfect information and the resulting effects on prices (Weisbuch et al., 2000; Kirman and Vriend, 2001; Moulet and Rouchier, 2007).

This paper aims to contribute towards this sparse literature. Having the possibility to observe the Pescara[1] wholesale fish market (MIPE, that in Italian stands for Mercato Ittico PEscara), we build an agent based simulation model to find out which buyers' behavior is able to replicate its price dynamics.

The present work differentiates itself from the cited literature on perishable products because the market under investigation has a centralized structure organized as a Dutch auction instead of being based on bilateral trading. From this point of view the paper also contributes to understand the anomalies of price dynamics highlighted by the empirical papers on sequential auctions for art, wine and jewelry (Ashenfelter and Genovese, 1992).

The paper is organized as follows. After the description of the market (Sect. 2.2), we model the buyers' bidding behavior using basic economic considerations (Sect. 2.3). In Sect. 2.4 we show the results of simulated market sessions under different conditions. In this part of the paper we also initialize the simulation using real data so that we are able to validate the buyers' behavior we formed in Sect. 2.3. Section 2.5 presents some comments and concluding remarks.

2.2 Market Description

The MIPE is organized as two simultaneous Dutch auctions. We'll refer to the three types of agents operating on the market as sellers, buyers and auctioneers. Sellers are the trawler owners who bring the fish to the market after catching it in the nearby sea. Buyers buy the fish to sell it in turn to the final consumers, to supermarkets or to transform it (cooking or making ready to cook products). Finally, the auctioneers direct the transactions and decide the initial price for each case of fish (products are arranged in cases of 4–5 kilos).

Description of a typical session. Before the beginning of the auction, sellers are randomly selected and the first two are each assigned to one of the two conveyor belts crossing the market hall. When the belts start moving, the selected sellers put the cases on one end of the assigned belt and the cases move slowly towards the other end. Each case still unsold stops on the automatic weighing machines located in the middle of the conveyor belts. The auctioneers move the fish to let the buyers also see the fishes lying in the bottom of the case; they often take a fish from the case and show it to the attenders. The auctioneers communicate to the cabin operators

[1] Pescara is a medium size Italian city situated approximately in the middle of the Italian east coast.

Table 2.1 Absolute frequency (#) and percentage of buyers and sellers by size

10^3 kg. →		0–5	5–10	.	.	.	.	.	.	.	.	50–55	.	.	.	.	.	.	.	.	95–100	>100
buyers	#	53	25	13	15	9	5	4	1	1	1	0	0	0	1	0	0	1	0	0	0	3
	%	40	19	10	11	7	4	3	.8	.8	.8	0	0	0	.8	0	0	.8	0	0	0	2
sellers	#	8	2	6	7	5	2	3	3	6	6	0	1	1	1	3	2	1	1	0	1	0
	%	14	3	10	12	8	3	5	5	10	10	0	2	2	2	5	3	2	2	0	2	0

(each of the two auctioneers stands by a cabin where a person is managing the market information system on a computer) the fish variety and the initial price. The cabin operators manage the displaying of the name of the seller, the fish variety and the initial price on the two large screens. The price counter starts going down on the display. Buyers are equipped with a remote control which is able to stop the counter. The buyer who stops the counter gets the case and, if the cases which follow contain the same variety, the buyer is asked by the auctioneer how many cases he would buy for the same price. Whenever a seller finishes, the next seller starts loading onto the free conveyor belt. As a result, all the cases of the same seller run sequentially on the same conveyor belt.

Data set description. The dataset contains data from October 3rd, 2005 to May 18th, 2007 and records for every transaction: (1) date and time; (2) product, buyer and seller's identification number; (3) the quantity (weight in kilos), price per kilo and number of cases; (4) the conveyor belt identification number. In the analyzed period, 59 sellers and 132 buyers operate 261,055 transactions on 72 varieties for a total amount of 1,974,316 kg of fish exchanged. The market operators meet once or twice in the working-days and extraordinarily on Saturday.

Let us discuss briefly the features of buyers and sellers operating in this market. Table 2.1 reports the frequency of them by classes of 5,000 kg of fish exchanged.

From the table it is possible to see how buyers are concentrated in the first two classes and are in general smaller than sellers. However a number of them have a significant size.[2] Sellers are less concentrated and there is no prevalent size among them.

As in the following discussion we concentrate on buyers, it would be useful to illustrate how they distribute according to the type of activity and the distance of their working place from the market. A bi-variate frequency distribution is presented in Table 2.2.

2.3 Modeling the Buyers' Bidding Behavior

Buyers in wholesale markets are in turn sellers in the retail market. To model their bidding behavior we have to take into account what happens in the final market. Having a well defined geographical zone, retailers move in an imperfect competition

[2] The three large buyers (that fall into the residual class) bought 265,749.10, 143,163.20 and 125,117.50 kg, respectively.

Table 2.2 Absolute and relative frequency of buyers by activity and distance of the working place from the market. Fish shops include points of sale in supermarkets and indoor markets; restaurant include catering; a number of fish shops make ready to cook products

activity → distance (km.) ↓	peddler	fish shop	restaurant	wholesaler	total
(0–15]	41 (0.311)	28 (0.212)	11 (0.083)	6 (0.045)	86 (0.65)
(15–45]	8 (0.061)	9 (0.068)	5 (0.038)	3 (0.023)	25 (0.19)
(45–85]	2 (0.016)	2 (0.015)	0	0	4 (0.03)
(85–135]	9 (0.068)	3 (0.023)	0	1 (0.007)	13 (0.10)
(135–)	1 (0.007)	1 (0.007)	1 (0.007)	1 (0.007)	4 (0.03)
total	61 (0.46)	43 (0.33)	17 (0.13)	11 (0.08)	132 (1)

context (Greenhut et al., 1987). Consequently, each of them faces a downward sloping demand curve, but his situation differs from that of each other by the location and the size of the market. The latter elements are important for understanding and modeling the bidding strategy during the auction. The location of the market for example imposes time constrains on retailers holding their activity far away from the port: they should attend the initial part of the auction to leave early. The size of the market is of course important to establish the maximum amount the buyer can buy, but also the minimum amount he must buy to hold his customer relationships. Finally it is worth pointing out that in the retail markets we are talking about, buyer-seller interactions are based on bilateral bargaining and prices are posted.

According to the above discussion the bidding behavior comes out from a process where buyers attempt to maximize the profit under the quantities and time constraints. Let us investigate it more formally.

Let p_i^s be the posted selling price obtained form buyer's i inverse demand, $\langle p^b \rangle_i$ the average price the buyer pays on the wholesale market and q_i the quantity. Of course p_i^s depends negatively on q_i ($p_i^s = p_i(q_i)$).

First of all the buyer tries to maximize profit (π_i), but differently from the textbook case he has the possibility of managing the average buying price during the auction. The mathematical problem is

$$\max_{(q_i, \langle p^b \rangle_i) \in R^{+2}} \pi_i = p_i^s q_i - \langle p^b \rangle_i q_i = p_i(q_i) q_i - \langle p^b \rangle_i q_i .$$

Given the positivity constraints, the optimal solution is $\langle p^b \rangle_i = 0$ and q_i satisfies $\frac{dp_i}{dq_i} q_i + p_i(q_i) = 0$. This gives us a first insight into the final goal of a buyer in the auction: when he leaves the market his $\langle p^b \rangle_i$ and q_i should be as close as possible to the optimal values.

Having established the final goal of a buyer, the rest of this section aims to provide a basic[3] framework for assessing how buyers manage to achieve the final goal, that is we analyze the bidding dynamics. The importance of timing pushes us

[3] Of course the question can be treated in a more detailed and complicated way, but our goal here is to grasp the essential behavior of buyers. Complications are left for future studies.

towards a more detailed notation. Let us identify a case of fish with $z \in \{1,2,....,Z\}$ and with t_z the time it appears on the market. The pair $(\langle p^b \rangle_{i,j,z}, q_{i,j,z})$ informs about the situation of buyer i when case z appears given that he has already bought j cases of fish. Using these variables one can compute a buyer's profit in a point in time $(\pi_{i,j,z})$. Our first step towards the bidding dynamics is to identify two upper bidding values (details are given in the appendix 2.6.1): the first one is obtained imposing the condition $\pi_{i,1,z} = 0$ and the second one imposing $\pi_{i,j,z} = \pi_{i,j+1,z}$. We combine them to get the dynamics of the upper bidding threshold:

$$\bar{p}^b_{i,j+1,z} = \begin{cases} p^s_{i,j+1,z} & \text{if } j = 0 \\ \left[p^s_{i,j+1,z} q_{i,j+1,z} - p^s_{i,j,z} q_{i,j,z} \right] \frac{1}{q_{i,j+1,z} - q_{i,j,z}} & \text{if } j > 0 \end{cases}. \tag{2.1}$$

The above calculations inform the buyer about what he has to do in order to ensure his achieved profit doesn't decrease. Furthermore, the solution of the above discussed maximization problem tells us that his goal when bidding for a new case is to lower the average buying price as much as possible. This can be obtained by bidding a lower value than the threshold, but on the other hand this fact lowers the chance of realizing this improvement by getting the case. At this stage, the constraints play a crucial role. In this model, when the constraints increase their binding power the buyer bids at a closer and closer level to the upper threshold because it augments the chance of realizing a profit improvement even if this improvement is small. To use an amusing expression, when the constraints are highly binding a buyer would think: "a bird in the hand is worth two in the bush."

To model this step by step search for improvements we introduce the constraints into the analytical formulation in the following way. Each buyer is characterized by four variables: the arrival time (a_i), the departure time (d_i), the minimum $(q_{i,\min})$ and the maximum quantity $(q_{i,\max})$. Provided that $a_i < t_z < d_i$, we let the bid of buyer i for transaction z evolve as

$$b_{i,z} = \begin{cases} \bar{p}^b_{i,j+1,z} \left(1 - \beta \left[\frac{d_i - t_z}{\alpha(q_{i,\min} - q_{i,j,z}) + (q_{i,\max} - q_{i,\min})} \right] \right) & \text{if} \quad q_{i,j,z} < q_{i,\min} \\ \bar{p}^b_{i,j+1,z} \left(1 - \beta \left[\frac{d_i - t_z}{(q_{i,\max} - q_{i,j,z})} \right] \right) & \text{if} \quad q_{i,\min} \le q_{i,j,z} < q_{i,\max} \end{cases}$$

where α and β are parameters.

The ratio in square brackets gives the relative strength of the constraints. Both of them decrease during the auction. The figure in the numerator (the amount of left) decreases at a constant speed while that in the denominator (the remaining quantity to buy) decreases whenever the buyer gets the fish. If the two speeds are the same, the ratio stays constant and the same happens to the bid that remains lower than the threshold. This gap is motivated by the attempt of the buyer to achieve the final goal at the end of the auction (the lowest possible average buying price) and its size depends on the β parameter. If fish accumulates at a higher speed than the time which passes, the buyer is in a good position and he exploits the chance to increase his profit decreasing the bid, while if the buyer feels in shortage of fish he increases the bid to improve his probability of getting additional fish. In the given formulation we allow the possibility that the buyer behaves differently depending on whether he has reached the quantity that permits him to hold his customer relationships. α represents the importance the buyer gives to the customer relationships (note that the first equation reduces to the second if $\alpha = 1$).

We close the discussion with the natural conclusion: the highest bid wins, so that the price at which case z is sold is

$$p_z = \max(b_{i,z}).$$

2.4 Simulations and Validation

In our simulation we simplify the model assuming that cases are identical in quality and weight; this allows us to formulate the model in terms of the number of cases instead of in the weight of fish (that is we replace the variable q with j). Moreover, we re-scale time to associate to each event a time index between 0 (the beginning of the auction) and 1 (the end of the auction). To set up the initial conditions we take as a benchmark the exchanges that took place on hake the 24th April 2007.[4] In this auction 58 buyers are present and 236 cases are exchanged. We ran simulations under three different conditions to see how the simulated price dynamics react to changes in the relevant variables. The simulation settings are reported in appendix 2.6.2. In the first exercise we set a_i, d_i, t_z, $j_{i.\max}$ randomly and $j_{i.\min}$ as a fraction of $j_{i.\max}$. Simulations under these conditions allow us to investigate how the price dynamics is affected by each variable. As an example we show in Fig. 2.1 how the price dynamics react to changes in the number of available cases. In the second exercise we set a_i, d_i, $j_{i.\max}$ according to the real data and t_z randomly. We report again on the effect of the availability of fish on price dynamics (see Fig. 2.2). In the third exercise all the previous mentioned variables are loaded using real data.

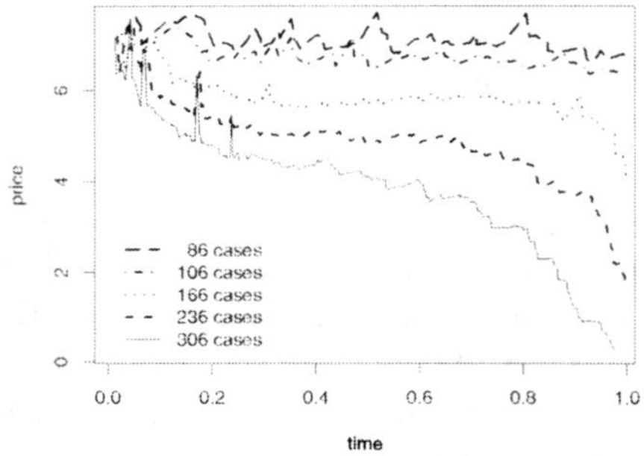

Fig. 2.1 Effect of the available quantity on price dynamics with random agents' characteristics and random process for the arrival of cases

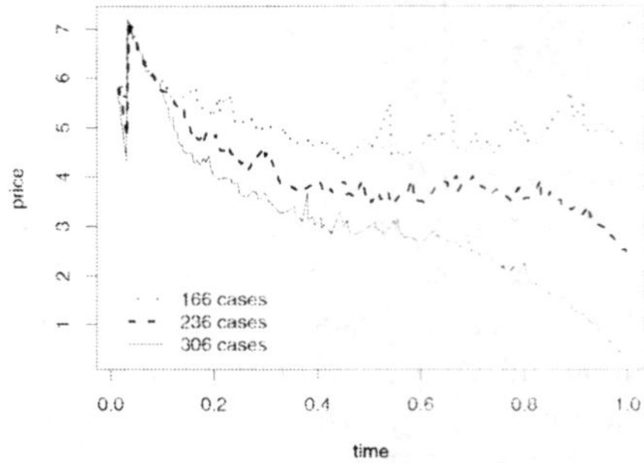

Fig. 2.2 Effect of the available quantity on price dynamics with agents' characteristics from real data and random process for the arrival of cases

[4] We carefully select this auction because a large amount of fish was exchanged and cases of hake seem to be of the same quality because prices have no large differences in the two conveyor belts.

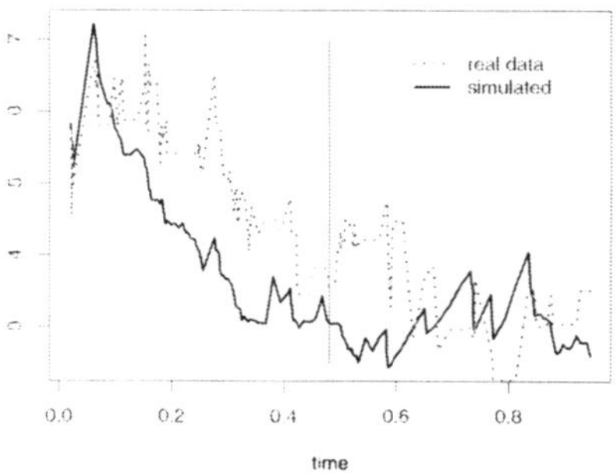

Fig. 2.3 Comparison between real and simulated price dynamics

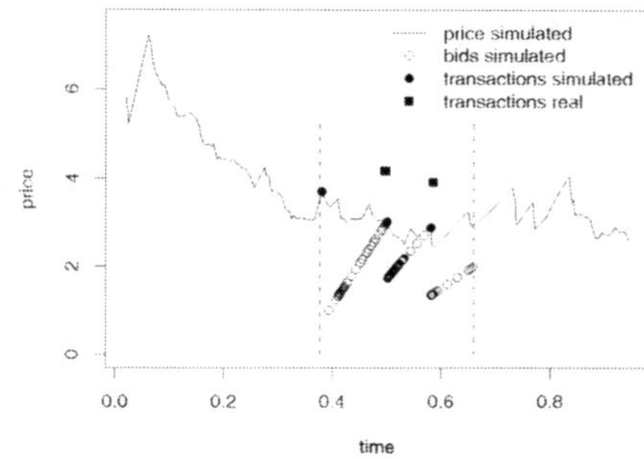

Fig. 2.4 A buyer's bidding dynamics

Table 2.3 Comparison between the number of cases bought by each buyer obtained from real data and those resulting from the simulation

buyer	1	2	.	.	.	.	.	.	.	.	.	.	.	.n	.	.	.	.	.	.	.	.	.	.	.	.	.	28	29
real data	3	1	4	2	2	3	3	1	3	2	1	4	15	3	2	2	1	1	1	1	1	1	1	8	1	1	3	2	2
simulated	3	1	4	2	2	3	3	1	3	2	1	4	15	3	2	2	1	1	1	1	1	1	1	8	1	1	3	2	2
difference	0	0	0	0	0	0	0	0	0	0	0	0	0	0	0	0	0	0	0	0	0	0	0	0	0	0	0	0	0

buyer	30	31	.	.	.	.	.	.	.	.	40	.	.	.	.	.	.	.	.	.	.	.	.	.	.	55	.	57	58
real data	1	1	5	22	3	12	1	2	1	2	**5**	1	3	2	2	1	7	10	1	1	6	1	1	3	2	**58**	3	3	1
simulated	1	1	5	22	3	12	1	2	1	2	**6**	1	3	2	2	1	7	10	1	1	6	1	1	3	2	**57**	3	3	1
difference	0	0	0	0	0	0	0	0	0	0	**−1**	0	0	0	0	0	0	0	0	0	0	0	0	0	0	**1**	0	0	0

In the last exercise the natural term of comparison is represented by real data. We concentrate our attention on this comparison and we try to validate the simulation result with the following tests. First of all, we compare the real and the simulated price dynamics (Fig. 2.3); although the simulated series starts to fall before and it remains below the real one for a large part of the auction, its dynamics mimic the real series in several episodes (especially before the vertical line). Secondly, we compare the number of cases bought by each buyer in real data and in simulations (see Table 2.3); the simulations fail to replicate this data on two occasions: buyer 40 buys 1 case more in the simulation than in reality (6 instead of 5) and buyer 55 buys one case less (57 instead of 58).

Finally, we compare the real and simulated average price of buyers by testing the following linear regression:

$$\langle p^b \rangle_i^{\mathrm{real}} = \eta + \xi \langle p^b \rangle_i^{\mathrm{simulated}} .$$

The results are shown in Table 2.4.

To give an idea of the individual bidding behavior generated by the model we report in Fig. 2.4 the situation for buyer 42 (the vertical dashed lines signal his arrival and departure times); note how the simulation replicates the timing of two of the three transactions of this buyer.

Table 2.4 Result of the linear regression between real and simulated average prices

	Estimate	Std. Error	t value	$Pr(>\|t\|)$
η	3.1928	0.7475	4.272	7.6e-05 ***
ξ	0.5042	0.2073	2.432	0.0182 **

Signif. codes: ***=0.01, **= 0.05.

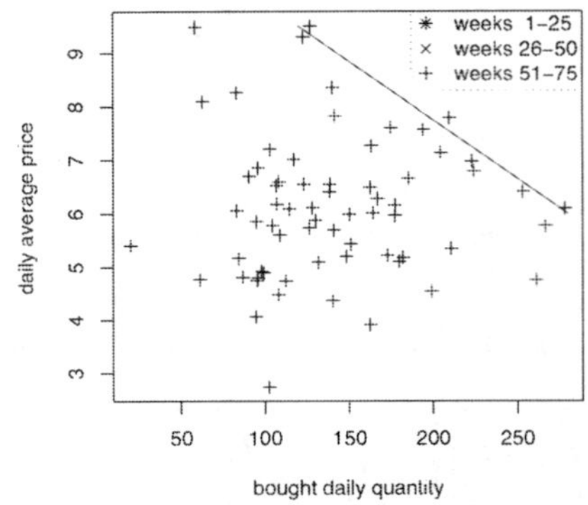

Fig. 2.5 Bought quantities and average prices of a buyer in the Tuesday morning auction

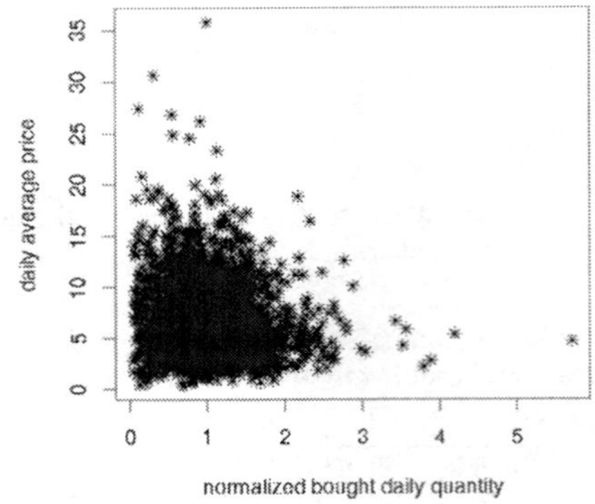

Fig. 2.6 Bought quantities and average prices of all the buyer in the Tuesday morning auction

Until now we have concentrated on price dynamics of a particular day. A more general test of the proposed model can be obtained by collecting the quantity and the average price obtained by buyers at the end of each auction. According to the model, in order to maximize profit, the pair $(q_i, \langle p^b \rangle_i)$ should be found on the marginal revenue curve. We noted above that the competition among buyers during the auction's progress could prevent the achievement of this theoretical position. It is unlikely however to find points in the north–east area of the marginal revenue curve because to arrive at this area a profit reduction should have been accepted by the buyer. In the end, the $(q_i, \langle p^b \rangle_i)$ pairs obtained by a buyer in the various auctions should be found on the south–west of a downward sloping line. Plotting this data, the obtained shape should be similar to a triangle and the closer the cluster of points is to a line, the higher the buyer's ability to maximize profit, i.e. to be on his marginal revenue curve. This reasoning provides us with a possible way to estimate a buyer's marginal revenue curve: if its position doesn't change with time, it is given by the north–east border of the cluster of points. We carry out this more general test plotting the result obtained on all the daily transactions (that is pooling all the species) of each individual buyer. Figure 2.5 reports the plot of quantities and average prices obtained by a single buyer in the Tuesday morning auctions. On this graph, the line is an attempt to locate this buyer's marginal revenue curve, and data from three time periods are reported with different symbols to check for time stability of the curve. Finally, to avoid the arbitrary nature of the buyer choice, we pool the data from all the buyers. To take into account differences in size, quantities are normalized dividing them by the average quantity bought by the buyer. Figure 2.6 shows the results of this exercise for the Tuesday morning auction. The expected triangular shape of the cluster of points can be detected. No significant changes in the shape of the cluster are observed plotting the data for the other weekday auctions.

2.5 Discussion and Conclusions

Understanding price dynamics of real estate markets is a challenging endeavor that receives little attention because of the lack of data. In this paper we attempt to make a contribution to this limited literature using data from the Pescara wholesale fish market. In the market we analyze, elements like the buyers' arrival and departure during the auction unwinding and the timing of products' appearance are important factors that affect the price dynamics.

A feature of price dynamics generated by sequential auctions that is puzzling the economic profession is their downward trend. This phenomenon is known as the "decreasing price anomaly" or "afternoon effect" (Ashenfelter and Genovese, 1992). This tendency is also present in MIPE, but according to our empirical investigations, the beginning and the final part of an auction often present non decreasing prices.

We model buyers in a centralized market with sequential appearance of products as agents that adaptively look for step by step improvement of their situation. The adaptive behavior is a potential explanation for a declining price. In fact, buyers in wholesale markets are in turn sellers in the retail markets and they face a downward sloping demand curve in the final market. The sequentiality of the auction implies that buyers cannot directly achieve their optimal point but they have to drive the average buying price down their marginal revenue curve over time. Our investigation also shows that this evolutionary process often ends up in a suboptimal position (in the south–west area of the marginal revenue curve) where notwithstanding buyers realize positive profits.

Our simulations, especially when the initial conditions are set using real data, give results that are compatible with reality. The changes in the degree of competition involved by the buyers' turnover are important to explain the price dynamics especially at the beginning and for the final part of the auction.

In our empirical investigations, complications arise because the product sold recorded under the same item eventually have qualitative differences that cannot be identified by the researcher. To overcome these difficulties, we carry out our work under rather restrictive conditions (we restrict the analysis to one species in one auction). However the existence of jumps in simulated prices signals that the price volatility observed in real data is not necessarily due to unobserved qualitative differences. This opens the way to a more general modeling strategy.

2.6 Appendix

2.6.1 The Bidding Threshold

Let us start from the profit for the first transaction:

$$\pi_{i,1,z} = p^s_{i,1,z} q_{i,1,z} - \langle p^b \rangle_{i,1,z} q_{i,1,z}.$$

The $\pi_{i,1,z} = 0$ condition brings us directly to the first threshold given in (2.1) once we note that for the first transaction $\langle p^b \rangle_{i,1,z} = p^b_{i,1,z}$.

For the subsequent transactions we require that the profit must not decrease. The highest bid is obtained from the condition $\pi_{i,j+1,z} = \pi_{i,j,z}$:

$$p^s_{i,j,z} q_{i,j,z} - \langle p^b \rangle_{i,j,z} q_{i,j,z} = p^s_{i,j+1,z} q_{i,j+1,z} - \langle p^b \rangle_{i,j+1,z} q_{i,j+1,z}.$$

Solving $\langle p^b \rangle_{i,j+1,z}$ we have

$$\langle p^b \rangle_{i,j+1,z} = \frac{1}{q_{i,j+1,z}} \left(p^s_{i,j+1,z} q_{i,j+1,z} - p^s_{i,j,z} q_{i,j,z} + \langle p^b \rangle_{i,j,z} q_{i,j,z} \right)$$

but

$$\langle p^b \rangle_{i,j+1,z} := \frac{1}{q_{i,j+1,z}} \left(\langle p^b \rangle_{i,j,z} q_{i,j,z} + (q_{i,j+1,z} - q_{i,j,z}) p^b_{i,j+1,z} \right)$$

equating them and solving $p^b_{i,j+1,z}$ one arrives at the second threshold given in (2.1).

Assuming identical weight (c) for cases we can substitute $q_{i,j,z}$ with $c j_{i,z}$ and we get the threshold we use in simulations:

$$\bar{p}^b_{i,j+1,z} = \left[p^s_{i,j+1,z} (j_{i,z} + 1) - p^s_{i,j,z} j_{i,z} \right].$$

2.6.2 Simulations Settings

In all the simulations we set $\alpha = 5$, $\beta = 25$ and the number of buyers to 58. We assume that buyers have a linear inverse demand function

$$p^s_{i,j,z} = \gamma - \frac{\gamma}{j_{i,\max}} j_{i,z}$$

and we set $\gamma = 8$.

Randomly generated values are obtained in the following way.

The arrival times are drawn from a beta distribution. This distribution supplies values between 0 and 1 and its shape can be controlled by two parameters. In particular we use $a_i \sim \beta(1, 30)$.

To establish the departure time we assume that each buyer attends at least a 10% of the auction duration. The remaining attendance time is represented by a variable u_i that is a realization of a uniform distribution $u_i \sim U(0,1)$:

$$d_i = \min(a_i + 0.1 + u_i, 1).$$

t_z is a Poisson process; to obtain the timing of cases when the total number of exchanged cases is Z, we draw Z numbers from an exponential distribution. Referring to them with τ_z, we compute

$$t_z = \frac{\sum_{w=1}^{z} \tau_w}{\sum_{w=1}^{Z} \tau_w}.$$

In the simulations, $j_{i,\max}$ is an integer random variable uniformly distributed between 1 and 30 and $j_{i,\min}$ is obtained rounding $0.2 j_{i,\max}$.

Acknowledgements We thank the Resal software staff which is collecting the data and delivered us a high quality dataset. The fish market coordinating staff, in particular Antonio Ciriaco, spent a considerable amount of time giving us additional information that are not included in the dataset.

References

W. B. Arthur, J. H. Holland, B. LeBaron, R. Palmer, and P. Tayler. Asset pricing under endogenous expectations in an artificial stock market. In W. B. Arthur, S. N. Durlauf, and D. A. Lane, editors, *The economy as an evolving complex system II*. Addison-Wesley, Reading, MA, 1997.

O. Ashenfelter and G. Genovese. Testing for price anomalies in real estate auctions. *American Economic Review*, 82:501–505, 1992.

K. Graddy. Testing for Imperfect Competition at the Fulton Fish Market. *The RAND Journal of Economics*, 26:75–92, 1995.

M. L. Greenhut, G. Norman, and C. Hung. *The Economics of Imperfect Competition: A Spatial Approach*. Cambridge University Press, Cambridge, 1987.

A. Kirman and A. Vignes. Price dispersion: Theoretical considerations and empirical evidence from the marseilles fish market. In K. G. Arrow, editor, *Issues in contemporary economics*. Macmillan, London, 1991.

A. Kirman and N. Vriend. Evolving market: An ace model of price dispersion and loyalty. *Journal of economics dynamics and control*, 25:459–502, 2001.

A. Kirman, R. Schulz, W. Härdle, and A. Werwatz. Transactions that did not happen and their influence on price. *Journal of Economic Behavior & Organization*, 56:567–591, 2005.

S. Moulet and J. Rouchier. The influence of seller learning and time constraints on sequential bargaining in an artificial perishable goods market. GREQAM working paper n. 2007-16, 2007.

G. Weisbuch, A. Kirman, and D. Herreiner. Market organisation and trading relationships. *The Economic Journal*, 110:411–436, 2000.

Chapter 3
Market Behavior Under Zero-Intelligence Trading and Price Awareness

Lucia Milone

Abstract This paper studies the consequences on market performance of different behavioral assumptions about agents' trading strategies (*pure zero-intelligence, nice behavior* and *greedy behavior*) and of pre-trade quote disclosure (*price awareness*). The investigation is conducted according to different criteria, such as efficiency, volume and price dispersion. We can summarize the main results as follow. *Nice* behavior performs better than *greedy* behavior with respect to all the performance criteria. Information disclosure increases allocative efficiency under the assumption of *nice* behavior but is not enough to achieve the same result with *greedy* traders. In fact, *greedy* traders perform better in the closed-book scenario in terms both of allocative efficiency and transaction volume. Furthermore, *nice* traders leads to the same volume of transaction in the open- and in the closed-book scenario, despite to a higher level of allocative efficiency in the former than in the latter; this can be explained (at least partially) by the lower number of intermarginal buyers and sellers that fail to trade under information disclosure. *Pure zero-intelligent* traders are usually overperformed by both the other kind of agents and both in the open- and in the closed-book scenario.

3.1 Introduction

Market protocol and trading behavior may influence economic performance. There is a growing literature that has focused on mechanism evaluation and design with the aim to try to isolate the effect of market rules, market environment and agent behavior on market performance.

Information disclosure is commonly recognized as a fundamental issue in the design of markets. Despite increasing research attention, there is still little consensus

L. Milone
Department of Economics, University of Venice, Cannaregio 873, 30121 Venice, Italy, e-mail: lucia.milone@unive.it

on how quantity and type of information available influence market behavior; results are mixed and no definitive answer has emerged.

In this paper, I will investigate the consequences of (partial) pre-trade quote disclosure on market performance according to different criteria, such as allocative efficiency, transaction volume and price dispersion. I follow the spirit of the seminal paper by Gode and Sunder (1993), commonly recognized as a benchmark in the study of the algorithms for continuous double auction markets. In order to keep the model as simple as possible, I replace human traders by Zero-Intelligence programs that submit random bids and offers following different specified rules of behavior.

In particular, my attempt is to try to find an answer to the following questions:

1. How does price awareness (and, in particular, pre-trade information disclosure) affect CDA market outcomes?
2. Do agents' differences in trading behavior matter? And, if yes, how do they influence performance of the markets in two different scenarios with and without information disclosure?

The *design* issue in (1) is addressed by comparing CDA outcomes in two different frameworks; in the closed-book scenario there is no public or private information accessible, in the open-book scenario private information on the outstanding ask and bid is made available (see Sect. 3.2.2). The issue in (2) deals with the *evaluation* of the role played by three behavioral rules that are described in detail in Sect. 3.2.1.

I present results from a simulation study that leads to the following main findings. Price awareness may affect CDA outcomes but its positive effect is strictly related to the trading strategies that characterize the agents' behavior; alone, it is not enough. Traders that behave more aggressively and/or more often as a truth-teller, submit offers in favor of the market and help to increase the total level of allocative efficiency achieved by the institution. *Pure zero-intelligent* traders are usually outperformed by agents that follow different strategies, independently from information available.

The paper is organized as follows. Section 3.2 presents the model. It focuses on description of the behavioral assumptions and illustrates the simulation setup. Section 3.3 lists the outcome variables and summarizes results. Section 3.4 concludes, offers interpretations and suggests future possible directions of research.

3.2 The Model

There is an economy with N traders; every agent i is initially randomly endowed with a zero (buyer) or a single unit (seller) of one good. Each buyer receives a private valuation v_i; similarly, each seller receives a private cost c_i. Both valuations and costs are drawn from the same uniform distribution on $[0, 1]$.

I analyze the performance of a continuous double auction sequential protocol, given its relevance in real contexts. The protocol is organized in trading sessions in which all agents participate. Each agent can exchange at most one unit of the good at a time. Price offers to buy and to sell are submitted in a randomly selected order. Two order books list asks and bids sorted from the lowest to the highest and from the highest to the lowest, respectively. The lowest (highest) offer in the ask (bid) queue is called *best ask* (*best bid*). As soon as one bidder offers a price that is equal or better to the lowest sell offer, the CDA matches these offers at a price equal to the best bid or best ask, depending on which of them come first; then, these orders are removed from the books and the agents become inactive for the rest of the trading session. The books are completely cleared at the end of each session, when all agents have placed their offer[1] and summary data from the period is stored.

3.2.1 Behavioral Assumptions

This investigation involves three different behavioral rules. *Pure Zero-Intelligence traders* (henceforth, ZI) submit random bids and asks drawn from a uniform distribution; due to incentive rationality constraints, they are allowed to offer bids below their valuations (i.e. on $[0, v_i]$) and asks above their costs (i.e. on $[c_i, 1]$); see Gode and Sunder (1993). The other two behavioral rules force traders to determine their offers according to the best bid and the best ask currently listed in the books[2] and the trade-off between 'exploit the most from transactions' and 'increase probability to trade'. Each trader i is endowed with a fixed private value (call it γ_i) randomly drawn from a uniform distribution on $[0, 1]$. It represents his propensity to stay on the book waiting for possible future better gains from trade[3] instead of submitting a higher bid and acting in favor of immediate transaction. Call b the best bid and a the best ask currently on the books; agents assume that the equilibrium price is contained in the interval $[b, a]$ and, as a consequence, they move from a pure ZI trading behavior and place an offer closer to a reference point appropriately chosen in that interval. The reference point is determined by the (private) propensity γ_i and is equal to $b + [(a - b)\gamma_i]$. *Nice traders* (henceforth, N) draw a pre-offer from a triangular distribution with minimum equal to zero, maximum equal to one and mode equal to the reference point. A buyer that has a pre-offer less than his value offers a price exactly equal to the pre-offer; if it is not the case, he bids his own valuation. *Greedy traders* (henceforth, G) drawn their offers from a triangular distribution over the support $[0, v_i]$, with minimum equal to zero, maximum equal to v_i and mode equal to the reference point if it is lower than the valuation or equal to v_i if it is not the case. Sellers' behaviors follow by symmetry.

Both these trading strategies do not respond to an intention to mimic human behavior. The aim, following in spirit the seminal paper of Gode and Sunder (1993),

[1] This choice follows the approach of LiCalzi and Pellizzari (2008, same volume) that model zero-intelligence trading without resampling; results described in Sect. 3.3 are coherent with their findings.

[2] If books are empty or there is no information about bids and asks currently listed, traders assume that the best bid equals zero and/or the best ask equals one (see Sect. 3.2.2).

[3] Obviously, trade must occur within the same period since at the end of each trading session books are cleared.

is to suggest and try to model a variation of the ZI trading in which traders continue to behave randomly but react in some way to beliefs on price or to information available on it (price awareness). The most notable difference between the two rules consists in the fact that *nice traders* want to represent agents that behave in favor of the market more frequently than *greedy traders*. In particular, *nice traders* bid more aggressively and more often as truth-tellers, offering their own valuation/cost.

3.2.2 Open and Closed Book Scenarios

In order to isolate the role played by price awareness, I study two scenarios. In one (*closed* book), no information, public or private, is available. In order to define their trading strategies, agents behave on the basis of the worst possible scenario, assuming $a = 1$ (best ask) and $b = 0$ (best bid). In the other (*open* book), I introduce to the model a market maker who communicates pre-trade private advice concerning the outstanding bid and ask to the trader who is in charge to submit his offer. Information disclosure has to be defined as "partial," since the source of information made available concerns only the best (lowest) ask and the best (highest) bid currently listed in the books; nothing is known about other asks and bids in the queue. Note that no formal differences in beliefs and behavior exist between the case in which there is no information at all and the case in which information is allowed but books are empty.

3.2.3 Experimental Design

The simulation setup is the following.

I initially ran computerized experiments for each of the three behavioral rules described in Sect. 3.2.1 and report results for 1,000 agents (N) and 50 independent trading sessions (henceforth, Experiment 1). Both the scenarios reported in Sect. 3.2.2 are considered ($1 + 2 \times 2 = 5$ simulations). At the beginning of the first trading session, for each agent an initial endowment (0 or 1), a valuation/cost for the good ($v_i/c_i \in [0, 1]$) and a propensity parameter ($\gamma_i \in [0, 1]$) are randomly drawn from a uniform distribution. These elements define the parameter set that remains constant across trading sessions and across simulations; in fact, it is re-initialized in the same way at the beginning of each round and for each simulation in order to favor comparison. Differences in results are simply driven by the random order in which offers are submitted, that is the only element in which trading sessions differ.

As a robustness check, I run a second experiment (henceforth, Experiment 2) using four different initial parameter sets. I will not report results in details since they are consistent with what found in Experiment 1. The initial parameter set and the random order in which buyers and sellers arrive make no practical difference for allocative efficiency, trading volume and price dispersion.

The simulations are run using C++. The statistical and graphical analysis of the data are made using R.

3.3 Results

I am interested in study CDA market protocol with respect to different performance criteria under three distinct assumptions on agents' behavior and with respect to two scenarios that differ in terms of information available. Section 3.3.1 describes the outcome variables; after that, results are provided.

3.3.1 Outcome Variables

Marshallian profits are given by the sum of the achievable payoffs in a competitive equilibrium.[4] The *equilibrium price* is defined by the intersection between demand and supply functions obtained sorting buyers' valuations from the highest to the lowest and sellers' costs from the lowest to the highest. Given the fact that stepwise supply and demand functions are used, I decide to define the equilibrium price as the midpoint in the (possible) range of the CE prices. Since valuations and costs are the same across periods and across simulations, these outcomes do not vary. When transaction occurs, buyer and seller profits ($v_i - p$ and $p - c_j$, respectively) are computed. *Allocative efficiency* is measured by the ratio of the gains from trade by all traders to the maximum total profit that could have been earned by all traders (i.e. marshallian profits); see Smith (1962). *Volume* is the number of transaction per trading session; it is normalized using CE Volume equal to one. *Average transaction price* is the mean transaction price per period; i.e. in each trading session, it is the ratio between the sum of transaction prices and the number of transactions.

As an example, Fig. 3.1 presents the endowments of the demand and supply function in one of the market used in Experiment 2. The random setup (N = 1,000, Trading Session = 50) results in 491 buyers and 508 sellers. The equilibrium price equals 0.513257. Marshallian Profits are equal to 123.112.

At the end of each trading session I also keep trace of the number of inter-marginal[5] buyers or sellers that fail to transact. As suggested in the model introduced by Zhan and Friedman (2007), I compute a measure of this specific source of inefficiency (called *VI-inefficiency*) given by the ratio between the loss of total surplus

[4] "*Through the Marshallian path, the buyer whose redemption value is the highest trades with the seller whose unit cost is the lowest; next, the two second ranked traders trade; and so forth,*" Zhan et al. (2002).

[5] Buyers (sellers) with values v_i greater (costs c_i lower) than the equilibrium price are called intermarginal traders. They are supposed to trade in CE.

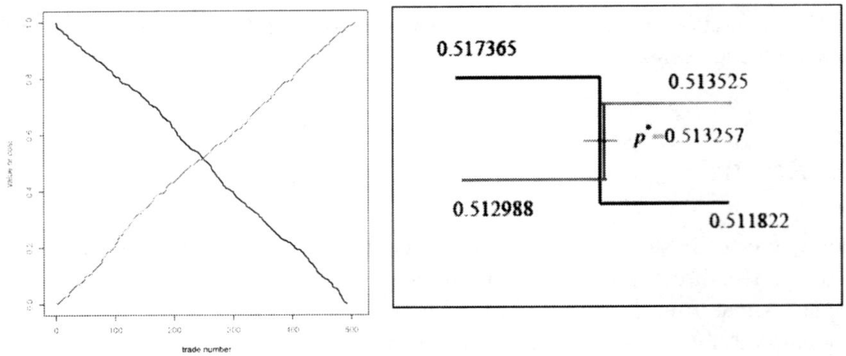

Fig. 3.1 Demand and supply functions in the CDA market – equilibrium price

over the marshallian profits.[6] It can be written as:

$$VI - inefficiency = \Big(\sum_{i \in IMBN} (v_i - p^*) + \sum_{j \in IMSN} (p^* - c_j) \Big) \Big/ \sum_{i,j=1}^{q^*} (v_i - c_j) \qquad (3.1)$$

where $IMBN$ and $IMSN$ are (respectively) the intermarginal buyers and sellers that fail to trade, q^* and p^* are the equilibrium quantity and price.

3.3.2 *Efficiency*

The left hand-side of Fig. 3.2 reports the allocative efficiency levels of the CDA market for 50 independent rounds. Each of the $1 + 2 \times 2$ possible combinations is identified by a different color: ZI is in black, N-closed is in red, N-open is in orange, G-closed is in blue and G-open is in green. The same coding applies for the right-hand side that reports data on average level of allocative efficiency in order to favor direct visual inspection.

The average allocative efficiency levels are 0.51 (ZI), 0.63 (N-closed), 0.65 (N-open), 0.56 (G-closed) and 0.53 (G-open).

For a statistical comparison of performance, I run a version of the Wilcoxon signed-rank test; significance of the differences is supported for any level of confidence and for any possible pair comparison (see Table 3.1).

The analysis of these results suggest two possible conclusions. First, *nice* behavior always performs better than a *greedy* behavior. Second, an open-book scenario performs better in terms of level of efficiency achieved only in a market with *nice-traders*. Pure zero-intelligence trading behavior is dominated by all the other possible scenarios.

[6] Zhan and Friedman (2007) demonstrate that the realized CDA surplus (i.e. allocative efficiency level achieved in the market) plus the loss due to VI-inefficiency and plus the loss related to trades that involve an extramarginal trader sum up to the marshallian profits, normalized to 1.

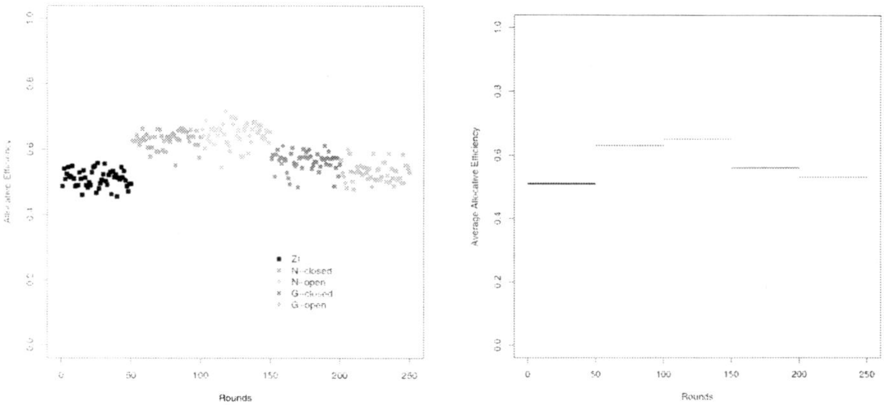

Fig. 3.2 Allocative efficiency

Table 3.1 Allocative efficiency

	ZI	*nice-closed*	*nice-open*	*greedy-closed*
nice-closed	0.0000*			
nice-open	0.0000*	0.0014*		
greedy-closed	0.0000*	0.0000*	0.0000*	
greedy-open	0.0013*	0.0000*	0.0000*	0.0000*

* *p-values* of the Wilcoxon test, *significant at the 0.05 level

3.3.3 Volume

The right-hand side of Fig. 3.3 reports the average transaction volume:[7] 0.477 (ZI), 0.795 (N-closed), 0.793 (N-open), 0.514 (G-closed) and 0.484 (G-open).

Results of the Wilcoxon test (Table 3.2) support significance of the differences for almost all the possible pair combinations; the most notable exception is the non-significance of the difference between the transaction volume in the closed and the open scenario for *nice* traders. *Nice* behavior still performs significantly better than *greedy* behavior, both in the open- and in the closed-book scenario. Coherent with what I found for allocative efficiency levels, *greedy* behavior performs better in a closed-book scenario. *Pure zero-intelligence* behavior leads to similar results in terms of total volume of transaction than *greedy* behavior under information disclosure.

These results suggest a further research question. If there is no significant difference in transaction volume between open and closed scenario under *nice* behavior, how can we justify significance of the difference in levels of allocative efficiency achieved?

[7] This measure is normalized using the number of transactions in CE: 240.

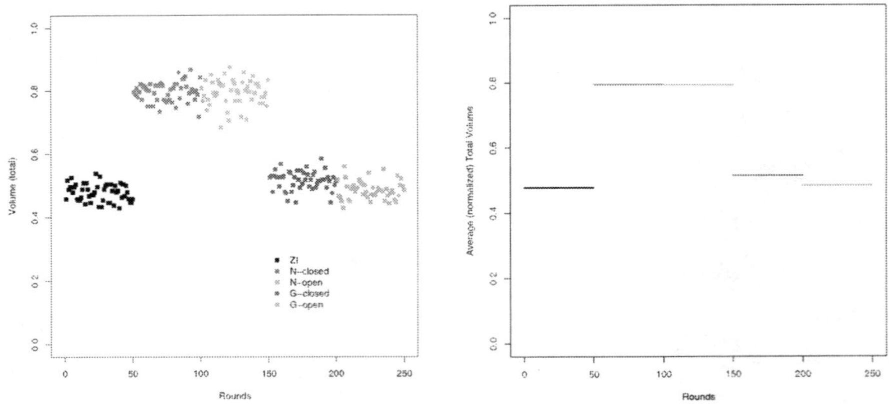

Fig. 3.3 Volume

Table 3.2 Total volume

	ZI	*nice-closed*	*nice-open*	*greedy-closed*
nice-closed	0.0000*			
nice-open	0.0000*	**1**		
greedy-closed	0.0000*	0.0000*	0.0000*	
greedy-open	**0.3287**	0.0000*	0.0000*	0.0000*

* *p-values* of the Wilcoxon test, *significant at the 0.05 level, **significant at the 0.1 level

In order to investigate this additional point at issue, I compute an alternative measure of inefficiency following the approach proposed by Zhan and Friedman (2007)

Average levels of VI-inefficiency are the following: 0.48 (ZI), 0.33 (N-closed), 0.32 (N-open), 0.43 (G-closed) and 0.46 (G-open) (Fig. 3.4).

VI-inefficiency for *nice* traders is lower (at a 1% level of significance) in the open book scenario and this fact can be used as a partial explanation for an higher level of allocative efficiency.

Note also that VI-inefficiency is lower under *nice* behavior than under *greedy* behavior, both in the open and in the closed book scenario. This result leads to the conclusion that trades between intermarginal traders are made easier by a more aggressive behavior in favor of the market and this supports the intuition that better performance of *nice* traders versus *greedy* traders is driven not only by an increase in transaction volume but also by a rise in the number of "fair" transactions (Table 3.3).[8]

[8] A robustness test has been conducted. *EMI-inefficiency* is defined by Zhan et al. (2007) as the total loss due to transactions that involve extramarginal traders; namely, it is equal to $p^* - v_i$ when an extramarginal buyer trades with an intramarginal seller and to $c_j - p^*$ when an extramarginal seller trades with an intramarginal buyer. Has been proved that (normalized) allocative efficiency in the CDA market = 1–VI*inefficiency* - EMI*inefficiency* (see footnote 6). As a consequence, when VI*inefficiency* lowers, the allocative efficiency of the market increases if EMI*inefficiency* decreases

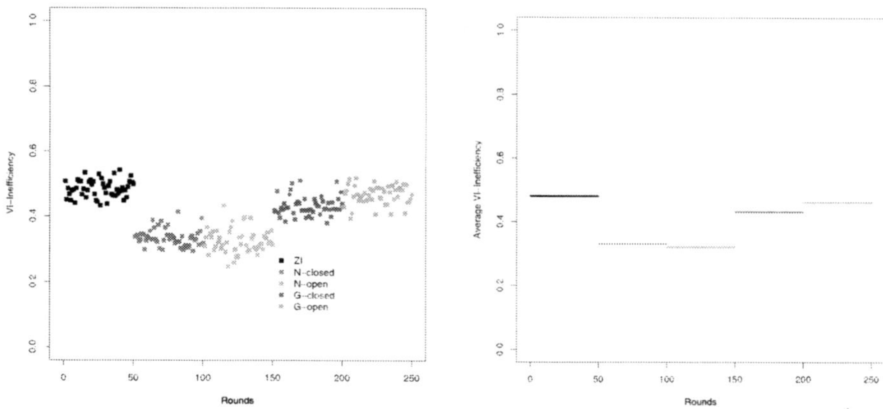

Fig. 3.4 VI-inefficiency

Table 3.3 VI-Inefficiency

	ZI	*nice-closed*	*nice-open*	*greedy-closed*
nice-closed	0.0000*			
nice-open	0.0000*	0.0854**		
greedy-closed	0.0000*	0.0000*	0.0000*	
greedy-open	0.0082*	0.0000*	0.0000*	0.0000*

* *p-values* of the Wilcoxon test, *significant at the 0.05 level, **significant at the 0.1 level

3.3.4 Transaction Prices

Figure 3.5 shows the average transaction prices time series for *nice* and *greedy* behavior in both an open- and a closed-book scenario, where the equilibrium price is normalized to 1.

Direct visual inspection shows that *nice* behavior performs better than *greedy* behavior both in the open and in the closed book scenario (see left-hand side of Fig. 3.5); i.e. offers are closer to the equilibrium price. Note that such a difference in performance is much greater under no information disclosure. At the same time, results seem to suggest that the open scenario performs better (in terms of price dispersion) both with *nice* and *greedy* traders (see right-hand side of Fig. 3.5); this unexpected result needs deeper investigation.

---

or remains constant; if it increases, the total allocative efficiency increases if $\triangle EMI < \triangle VI$, remains constant if $\triangle EMI = \triangle VI$ and decreases if $\triangle EMI > \triangle VI$. Situation when VI*inefficiency* increases follows by symmetry. Results show that EMI*inefficiency* is lower in an open book scenario both for *nice* and for *greedy* traders; significance in differences is supported by a Wilcoxon test for every level of confidence. Allocative efficiency for *nice* traders is higher in an open book scenario; allocative efficiency for *greedy* traders is higher in a closed book scenario since $\triangle EMI$ is significantly lower than $\triangle VI$. This confirms that "identity" of the traders matter.

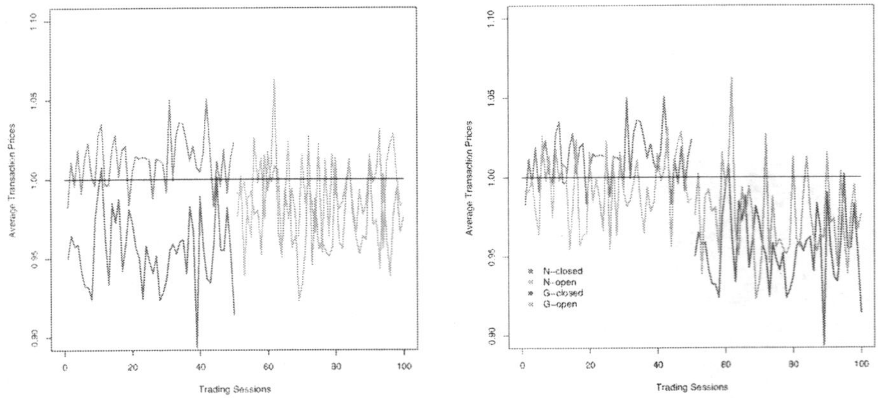

Fig. 3.5 Average (normalized) transaction prices (time series)

Even if market design seems to play a fundamental role in reducing dispersion, the fact that *nice* behavior continues to perform better independently from the information setup suggests that effects on price dispersion of trading strategies are stronger than the effects due to information available.

I performed a Wilcoxon signed-rank test on a standard measure of price dispersion; results confirm intuitions and differences are significant for any level of confidence and for every pairwise comparison.

3.4 Conclusions

This paper deals with two different issues. On the one hand it refers to a *design* problem: is or is not convenient to make information on outstanding bid and ask privately available? On the other hand it makes an attempt to evaluate the role played by different trading behaviors in different scenarios.

Simulation evidence leads to the following main conclusions.

Nice behavior performs better than *greedy* behavior with respect to all the performance criteria mentioned in the paper; namely, allocative efficiency, transaction volume, price dispersion (and VI-inefficiency).

Information disclosure increases allocative efficiency under the assumption of *nice* behavior but is not enough to achieve the same result with *greedy* traders. In fact, *greedy* traders perform better in the closed-book scenario in terms both of allocative efficiency and transaction volume. Furthermore, *nice* traders lead to the same volume of transaction in the open- and in the closed-book scenario, despite of a higher level of allocative efficiency in the former than in the latter; this can be explained (at least partially) by the lower number of intermarginal buyers and sellers that fail to trade under information disclosure.

Pure zero-intelligent traders are usually outperformed by both the other kind of agents and both in the open- and in the closed-book scenario.

An additional intuition that helps to interpret these results comes directly from the fact that a *nice* behavior works in favor of the market and increases probability to trade more than a *greedy* behavior. A *nice* trader who adopts a weakly dominated strategy (namely, bids his own valuation) with a higher and positive probability supports the risk to make zero profits but also favors the realization of a positive profit for other agents in the market; contrarily, a *greedy* trader that knows to have a low probability to trade prefers to make zero profits without trade and might offer a low price that prevents transactions.

This setup is organized over a single trading session, where all agents participate. At this stage there is no role for dynamics and each round works independently from each other. It could be interesting to investigate whether more realistic assumptions about agents' behavior affect results. My claim is that learning (on the basis of past history and past trades) in a dynamic context might significantly help in increasing efficiency moving trades closer to the equilibrium price (probably after an initial adjustment).

Another possible interesting direction for future research could be to try to investigate deeply the balance between nice and greedy behavior of the agents in the market and try to figure out which is the right proportion between them that permits to increase efficiency.

Finally, it might be interesting to compare outcomes of the CDA markets with different levels of information available both in terms of quantity and type.

References

D. Gode and S. Sunder. Allocative efficiency of markets with zero intelligence traders: Market as a partial substitute for individual rationality. *Journal of Political Economy*, 101:119–137, 1993.

M. LiCalzi and P. Pellizzari. The allocative effectiveness of market protocols under intelligent trading. In C. Bruun, editor, *Advances in Artificial Economics*, pages 17–29. Springer, Berlin, 2006.

M. LiCalzi and P. Pellizzari. Simple market protocols for efficient risk sharing. *Journal of Economic Dynamics and Control*, 31:3568–3590, 2007a.

M. LiCalzi and P. Pellizzari. Which market protocols facilitate fair trading? In A. Consiglio, editor, *Artificial Markets Modeling*, pages 81–97. Springer, Berlin, 2007b.

V.L. Smith. An experimental study of competitive market behavior. *Journal of Political Economy*, 70:11–137, 1962.

B. Toth, E. Scalas, J. Huber, and M. Kirchler. The value of information in a multi-agent market model : The luck of the uninformed. *The European physical journal B*, 55:115–120, 2007.

W. Zhan and D. Friedman. Markups in double auction markets. *Journal of Economic Dynamics and Control*, 31:2984–3005, 2007.

W. Zhan, J. Zhang, J. Yang, S. Wang, and K. K. Lai. k-ZI: A general zero-intelligence model in continuous double auction. *International Journal of Information Technology and Decision Making*, 1:673–691, 2002.

Part II
Evolution and Decision Making

Chapter 4
Evolutionary Switching between Forecasting Heuristics: An Explanation of an Asset-Pricing Experiment

Mikhail Anufriev and Cars Hommes

Abstract In this paper we propose an explanation of the findings of a recent laboratory market forecasting experiment. In the experiment the participants were asked to predict prices for 50 periods on the basis of past realizations. Three different aggregate outcomes were observed in an identical environment: slow monotonic price convergence, persistent price oscillations, and oscillatory dampened price fluctuations. Individual predictions exhibited a high degree of coordination, although the individual forecasts were not commonly known. To explain these findings we propose an evolutionary model of reinforcement learning over a set of simple forecasting heuristics. The key element of our model is the switching between heuristics on the basis of their past performance. Simulations show that such evolutionary learning can reproduce the qualitative patterns observed in the experiment.

4.1 Introduction

In social systems today's individual decisions depend upon expectations or beliefs about future developments. A financial market is a typical example of an *expectations feedback system*: market history shapes individual expectations which, in turn, determine current aggregate market behavior and so on. Clearly, the aggregate outcome depends then on the interaction of individual market forecasts and the ways individuals form market expectations. Traditional economic theory (see e.g. Muth, 1961; Lucas and Prescott, 1971) assumes that all individuals have rational expectations. In a rational world individual expectations coincide, on average, with market realizations, and markets are efficient with prices fully reflecting economic fundamentals.

M. Anufriev and C. Hommes
CeNDEF, University of Amsterdam, Roetersstraat 11, NL-1018 WB, Amsterdam, the Netherlands,
e-mail: [M.Anufriev,C.H.Hommes]@uva.nl

Already in the 1950's Simon (1957) observed that rationality imposes unrealistically strong informational and computational requirements upon individual behavior. Laboratory experiments discussed at length in Kahneman (2003) and Tversky and Kahneman (1974) indeed showed that individual decisions under uncertainty are at odds with perfectly rational behavior, and can be much better described by simple *heuristics*, which sometimes may lead to persistent biases. Another problem with rational expectations models is that they cannot explain many regularities observed in real markets, such as excess volatility and persistent deviations from rational equilibrium. Many economists argue nowadays that it is more reasonable to model individuals as *boundedly rational*, using simple rules of thumb in decision making.

Laboratory experiments with human subjects and controlled economic fundamentals are well suited to study how individuals form expectations and how their interaction shapes aggregate market behavior. But the results from laboratory experiments are mixed. Early experiments by Smith (1962) show convergence to equilibrium, while more recent asset pricing experiments exhibit deviations from equilibrium with persistent bubbles and crashes (Smith et al., 1988; Hommes et al., 2005). As Duffy stresses in his recent review (Duffy, 2008), a clear explanation of these different market phenomena is still lacking.

In recent learning to forecast experiments, described at length in Hommes et al. (2005), qualitatively different aggregate outcomes have been observed in the same experimental setting. In a stationary environment participants, during a number of periods, had to predict the price of a risky asset (say a stock) having knowledge of the fundamental parameters and previous price realizations, but without knowing the forecasts of others. If all agents would behave rationally or learn to behave rationally, the market price would quickly converge to a constant fundamental value. While in some groups convergence did happen, in other groups prices persistently fluctuated. What was even more striking, is that in all groups individuals were able to coordinate their forecasts. In this paper we present a simple model based on *evolutionary selection of simple heuristics* explaining how coordination of individual forecasts can emerge and, ultimately, enforce the different aggregate market outcomes. Simulations suggest that this model can explain *different* time series patterns in the *same* laboratory experiments.

The organization of the paper is as follows. Section 4.2 describes the forecasting experiment in detail, discusses the experimental results and provides a motivation for the model with heterogeneous expectations. Section 4.3 explains the main assumptions of the model. Numerical simulations of the model are presented and discussed in Sect. 4.4. Section 4.5 concludes.

4.2 Laboratory Experiment

A number of sessions of a computerized learning to forecast experiment have been performed in the CREED laboratory at the University of Amsterdam; see Hommes et al. (2005) for a detailed description. In each session six participants had to predict the price of an asset and have been rewarded according to the accuracy of their predictions. The participants were told that they are advisors to a pension fund and that this pension fund can invest money either in a risk free asset with interest rate r per period or in shares of an infinitely lived risky asset. In each period the risky asset pays uncertain dividends which is a random variable, independent identically distributed (IID), with mean $\bar{y}$. The price of the risky asset, p_t, is determined by a market clearing equation on the basis of the investment strategies of the pension fund. The exact functional form of the strategies and the equilibrium equation were unknown to the participants, but they were informed that the higher their forecast is, the larger will be the demand for the risky asset of the pension fund. Participants also knew the values of the parameters $r = 0.05$ and $\bar{y} = 3$, and therefore had enough information to compute the *rational fundamental price* (i.e. the discounted sum of the expected future dividend stream) of the risky asset $p^f = \bar{y}/r = 60$.

Every session of the experiment lasted 51 periods. In every period each of the 6 participants provided a two period ahead forecast for the price of the risky asset. The information participants knew consisted of past prices (up to two lags) of the risky asset and their own past predictions (up to one lag). Participants did not know the predictions of others. When all 6 predictions for the price in period $t+1$ have been submitted, the current market clearing price was computed according to a standard model of asset pricing, see e.g. Brock and Hommes (1998):

$$p_t = \frac{1}{1+r}\left((1-n_t)\bar{p}^e_{t+1} + n_t p^f + \bar{y} + \varepsilon_t\right), \tag{4.1}$$

where $\bar{p}^e_{t+1}$ denotes an (equally weighted) average of the 6 individual forecasts, ε_t is a stochastic term representing small demand and supply shocks, and n_t stands for a small fraction of "robot" traders who always submit a fundamental forecast p^f. These robot traders were introduced in the experiment as a "stabilizing force" to prevent the occurrence of large bubbles. The fraction of robot traders increased as the price moved away from its fundamental equilibrium level:

$$n_t = 1 - n\exp\left(-\frac{1}{200}\left|p_{t-1} - p^f\right|\right). \tag{4.2}$$

This mechanism reflects the feature that in real markets there is more agreement about over- or undervaluation when the deviation from the fundamental is large.

At the end of each period every participant was informed about the realized price and her earnings were defined by a quadratic scoring rule

$$e_{t,h} = \max\left(1300 - \frac{1300}{49}(p_t - p^e_{t,h})^2, 0\right) \tag{4.3}$$

with 1,300 points being equivalent to 0.5 euro. There were 7 sessions of the experiment. The stochastic shocks ε_t, drawn from normal distribution with mean 0 and standard deviation 0.5, were the same in all sessions.

4.2.1 Findings of the Experiment

The main findings of the experiment are as follows. First, three different price patterns were observed, see the upper parts of the six panels in Fig. 4.1. In groups 2 and 5 the price of the asset slowly converged, almost monotonically, to the fundamental price. In groups 1 and 6 the price oscillated around the fundamental price with an (almost) constant amplitude. In groups 4 and 7 large initial fluctuations were observed, dampening slowly towards the end of the experiment.[1] Different aggregate price behavior occurred despite the exact similarity between all seven sessions of the experiment.

Second, analysis of the individual price predictions reveals that during each session the participants were able to coordinate on a common prediction strategy, as illustrated in the lower parts of the panels in Fig. 4.1.

Finally, estimation of the individual predictions (based on the last 40 observations after a short learning phase) showed that participants had a tendency to use simple, linear forecasting rules. In the groups with monotonic convergence many participants formed their expectations adaptively, forecasting a weighted average of the last observed price and the last own forecast. In the other groups more complicated rules were observed, but most of the participants based their predictions on at most two recently observed prices. For example, participants often extrapolated price trends, multiplying the last price difference by some constant coefficient and then adding the result to the last price level.

4.2.2 Discussion

The three observed patterns in the experiment are very different from the predictions of the standard economic model with rational expectations (RE). In a world where all individuals are rational, the dynamics of prices should reflect the dynamics of the fundamentals (see Fama, 1970). In this experimental setting rational expectations yields the price dynamics shown in the left panel of Fig. 4.2, where the price deviations from the fundamental value $p^f = 60$ occur only due to the unpredictable demand/supply shocks ε_t. The inner frame shows the individual forecasting errors under RE. These forecasting errors can be compared with the errors in the experiment (in the right panel we show the example for treatment 7). Under RE the errors are smaller (ten orders of magnitude!) and do not have a lear structure, whereas errors in the experiments have significant auto-correlations.

One can argue that the RE model still provides useful predictions for the long run dynamics. Even if the economic agents acting in complex environment are not able to behave rationally from the very beginning, they will learn to do so. The learning

[1] Price dynamics in the remaining group 3 was more difficult to classify, somewhere between oscillations and convergence, probably due to a typing error by one of the participants in the middle of the experiment.

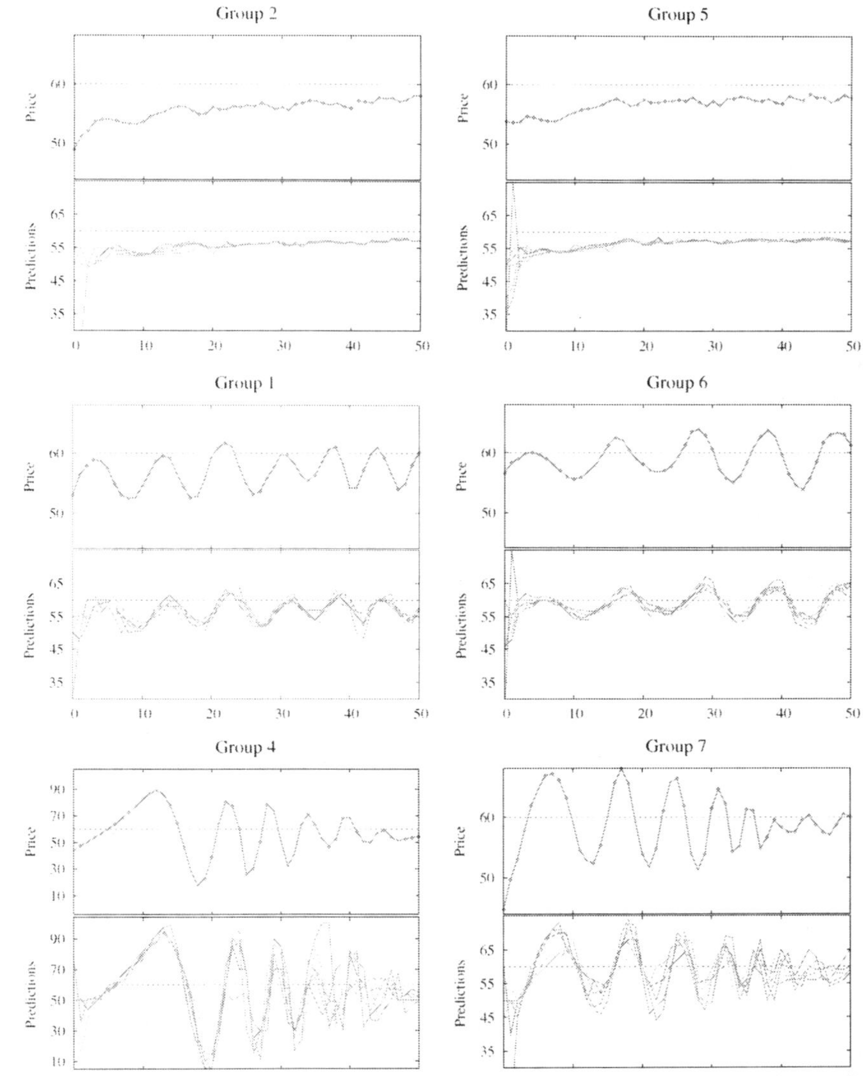

Fig. 4.1 Prices (*upper parts of the panels*) and individual predictions (*lower parts*) for 6 different treatments in the laboratory experiment with human subjects. Three different aggregate market outcomes were observed: slow monotonic convergence to equilibrium (*top panels*), permanent oscillations (*middle panels*) and oscillatory convergence (*bottom panels*). All treatments were characterized by a high degree of coordination of individual forecasts

literature inspired by Sargent (1993) and quickly developed in the last two decades puts emphasis on the conditions under which convergence to RE will occur.

From this perspective, the results reported above are somehow puzzling. In some groups the dynamics of individual predictions as well as aggregate dynamics moved

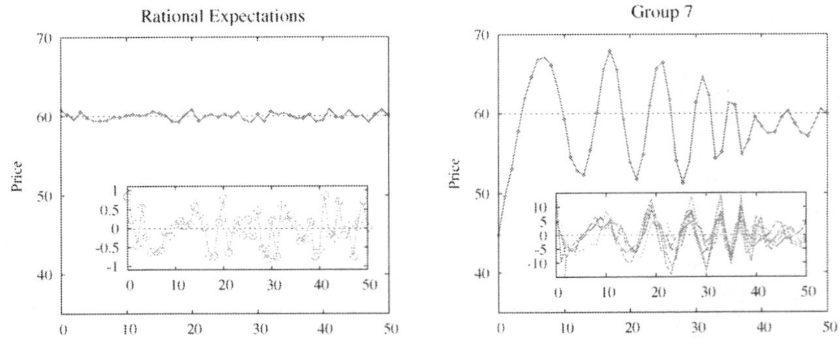

Fig. 4.2 Price and forecasting errors (*inner frame*) under Rational Expectations (*left panel*) and in the session 7 of the experiment

in the correct direction. One can conclude that in groups 2 and 5 and also, perhaps, in groups 4 and 7 convergence to the RE occurred towards the end of the experiment. On the other hand, no signs of such convergence were observed in groups 1 and 6. The existing models of the learning literature cannot explain these differences. Indeed, these models predict either convergence to or divergence from the RE equilibrium depending on the underlying economic law of motion, which was the same in all treatments.

The experiment can also be interpreted as a game with 6 participants, whose strategies are prediction sequences. It is easy to see that this game has a unique Nash equilibrium in pure strategies with all agents making fundamental prediction every period. The learning models from the game-theoretical literature then predict again either convergence to or divergence from the equilibrium, depending on the way people learn.

One important aspect, which is often ignored in existing models of learning, is heterogeneity in agents' expectations. In a recent review Hommes (2006) discusses models with heterogeneous expectations, which can explain different phenomena in financial markets. Heterogeneity can be also a key feature to explain the outcome of this experiment. Despite observed coordination on a common prediction strategy, many different rules generating different price patterns were estimated from the experiment. While every forecasting heuristic alone can generate only a certain type of dynamics, their combination together with learning over the set of heuristics may lead to different patterns.

4.3 Evolutionary Model

The above considerations led us to the following model. Assume that agents select forecasting rules from a population of simple heuristics. The choice of heuristic to use is governed by an *evolutionary selection* mechanism, so that more successful rules attract more followers. Performance of a heuristic is measured by accumulated (negative) squared prediction errors, in line with the payment incentives in the laboratory experiments.

4.3.1 Forecasting Heuristics

To keep our model as simple as possible, but rich enough to explain the different observed price patterns, we have chosen only 4 heuristics which are intuitively simple and were among the rules estimated on the individual forecasts in the experiment. A behavioral interpretation underlies each heuristic. The first heuristic is an *adaptive expectations* (ADA) rule, using a weighted average between the last observed market price and the last individual forecast. Note that at the moment when forecasts of price p_{t+1} are submitted, price p_t is still unknown (see (4.1)) and the last observed price is p_{t-1}. At the same time, the last own forecast $p^e_{t,1}$ is known when forecasting p_{t+1}. We have chosen the following ADA rule:

$$p^e_{t+1,1} = 0.65\, p_{t-1} + 0.35\, p^e_{t,1}\,. \tag{4.4}$$

The second and third heuristics are *trend following* rules extrapolating a weak or a strong trend, respectively. They simply predict the last observed *price level* plus a multiple of the last observed *price change*, and only differ in the magnitude of the extrapolation factor. In the case of *weak trend rule* (WTR) the factor is small and equal to 0.4, so that the rule is

$$p^e_{t+1,2} = p_{t-1} + 0.4\,(p_{t-1} - p_{t-2})\,. \tag{4.5}$$

The *strong trend rule* (STR) has a larger extrapolation factor 1.3 and is given by

$$p^e_{t+1,3} = p_{t-1} + 1.3\,(p_{t-1} - p_{t-2})\,. \tag{4.6}$$

The fourth heuristic is slightly more complicated. It combines an average prediction of the last observed price and an estimate of the long run equilibrium price level with an extrapolation of the last price change. More precisely, the rule is given by

$$p^e_{t+1,4} = 0.5\,(p^{av}_{t-1} + p_{t-1}) + (p_{t-1} - p_{t-2})\,, \tag{4.7}$$

where p^{av}_{t-1} is the sample average of all past prices, that is, $p^{av}_{t-1} = \Sigma_{j=0}^{t-1} p_j$. This rule is a *learning anchor and adjustment heuristic* (LAA), since it uses a (time varying) anchor or reference point, $0.5\,(p^{av}_{t-1} + p_{t-1})$, defined as an (equally weighted) average between the last observed price and the sample mean of all past prices, and extrapolates the last price change $(p_{t-1} - p_{t-n2})$ from there. The LAA rule has in fact been obtained from a linear forecasting rule $p^e = 30 + 1.5\,p_{t-1} - p_{t-2}$, used by some individuals in the experiment. In the experiment however, subjects did not know the fundamental price p^f explicitly, but were able to learn an anchor $0.5(p^f + p_{t-1})$ and extrapolate price changes from there.

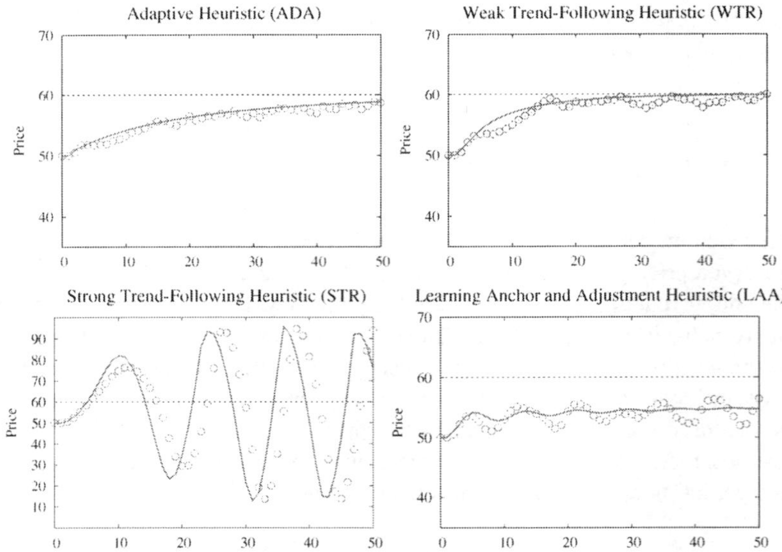

Fig. 4.3 Price dynamics in the models with homogeneous expectations corresponding to 4 different forecasting heuristics. The prices are simulated without an addition of noise terms (*lines*) and with an addition of the same noise sequence, ε_t, as was in the experiment (*circles*)

When one of these 4 heuristics is used alone, the dynamics given by (4.1) and (4.2) produces a certain pattern. These patterns are essentially the outcomes of 4 different models with *homogeneous* non-rational expectations. They are shown in Fig. 4.3 both for the "deterministic skeleton," i.e. when the pricing equation is not augmented by the noise realizations, and when the noise term ε_t is added to the pricing equation, as in (4.1). The model with the ADA or WTR heuristics leads to price convergence, the model with the STR heuristic generates wild (but non damping) price oscillations, while the model with the LAA heuristic (and experimental noise) produces price oscillations of constant amplitude (but around a low price level). Thus, each of the forecasting rules (4.4)–(4.7) can generate diverse dynamics but cannot explain all three qualitative patterns observed in the experiment, as well as coordination on a common prediction rule.

4.3.2 Evolutionary Switching

Which forecasting heuristics from the population should agents choose? Our simulation model is based upon evolutionary switching between the four forecasting heuristics, driven by the past relative performance of the heuristics. The performance of heuristic h, $1 \leq h \leq 4$, up to (and including) time period t is given by

$$U_{t,h} = -\left(p_t - p^e_{t,h}\right)^2 + \eta\, U_{t-1,h}\,. \tag{4.8}$$

The parameter $0 \leq \eta \leq 1$ measures relative weight agents give to past errors and thus represents their memory strength. When $\eta = 0$, only the performance of the last period plays a role in the updating of the shares assigned to the different rules. For $0 < \eta \leq 1$, all past prediction errors affect the heuristic's performance.

Given the performance measure, the weight assigned to rule h is updated according to a discrete choice model with *asynchronous updating*

$$n_{t,h} = \delta\, n_{t-1,h} + (1-\delta)\frac{\exp(\beta\, U_{t-1,h})}{Z_{t-1}}, \tag{4.9}$$

where $Z_{t-1} = \sum_{h=1}^{4} \exp(\beta\, U_{t-n1,h})$ is a normalization factor. There are two important parameters in (4.9). The parameter $0 \leq \delta \leq 1$ gives some persistence or inertia in the weight assigned to rule h, reflecting the fact that not all the participants are willing to update their rule in every period. Hence, δ may be interpreted as the fraction of individuals who stick to their previous strategy. In the extreme case $\delta = 1$, the initial weights assigned to the rules never change, no matter what their past performance is. If $0 \leq \delta < 1$, in each period a fraction $1-\delta$ of participants updates their rule according to the well known *discrete choice model* used for example in Brock and Hommes (1997) and Brock and Hommes (1998). The parameter $\beta \geq 0$ represents the intensity of choice measuring how sensitive individuals are to differences in strategy performance. The higher the intensity of choice β, the faster individuals will switch to more successful rules. In the extreme case $\beta = 0$, the fractions in (4.9) move to an equal distribution independent of their past performance. At the other extreme $\beta = \infty$, all agents who update their heuristic (i.e. a fraction $1-\delta$) switch to the most successful predictor.

Finally, the price p_t in our model is computed as

$$p_t = \frac{1}{1+r}\Big((1-n_t)\big(n_{t,1}\, p^e_{t+1,1} + \cdots + n_{t,4}\, p^e_{t+1,4}\big) + n_t\, p^f + \bar{y} + \varepsilon_t\Big), \tag{4.10}$$

where $p^e_{t+1,1}, \ldots, p^e_{t+1,4}$ are the predictions for period $t+1$ according to the 4 heuristics in (4.4–4.7), $n_{t,1}, \ldots, n_{t,4}$ are the fractions of the use of these heuristics described by (4.8–4.9). As before n_t stands for a small fraction of "robot" traders as in (4.2), $r(=0.05)$ is the risk free interest rate, $\bar{y}(=3)$ is the mean dividend, $p^f(=60)$ is the fundamental price and ε_t is the stochastic term representing small demand and supply shocks (taken to be the same as in the experiment).

4.3.3 Model Initialization

The model is initialized by two initial prices, p_0 and p_1, and initial weights $n_{1,h}$, $1 \leq h \leq 4$ (summing to 1; the initial share of robot traders $n_1 = 0$). Given p_0 and p_1, the heuristics forecasts can be computed and, using the initial weights of the heuristics, the price p_2 can be computed. In the next period, the forecasts of the heuristics are updated, the fraction of "robot" traders is computed, while the same initial weights $n_{1,h}$ for individual rules are used (past performance is not well defined yet in period 3). The price p_3 is computed and the initialization stage is finished. Starting from period 4 the evolution according to (4.10) is well defined: first the performance measure in (4.8) is updated, then, the new weights of the heuristics are computed according to (4.9) and finally a new price is determined by (4.10).

4.4 Simulations of the Model

Our evolutionary selection mechanism contains three parameters, β, η and δ, measuring, respectively (a) how sensitive individuals are with respect to differences in strategy performance, (b) how much relative weight they give to the most recent errors, and (c) how strongly an individual sticks to her previous strategy. In addition, five free initial values can be varied, which are two prices and the weights of three heuristics. We have performed numerous simulations and found that for a large range of parameters and initial values the heuristics switching model is capable to produce both persistent oscillating and converging patterns.[2] The model simulations in Fig. 4.4 illustrate that it is able to reproduce all three different qualitative patterns observed in the laboratory experiments, slow monotonic convergence to the fundamental price, persistent price oscillations, and dampened oscillatory movements. The left panels show simulated prices for the heuristics switching model, while the right panels demonstrate how the shares of the 4 heuristics are evolving. The parameters, initial prices and initial distribution over the heuristics used in the simulations are reported in Table 4.1.

To stress the independence of resulting price dynamics on the initial distribution of weights over heuristics, for all simulations reported in Fig. 4.4, we have chosen the same, uniform initial distribution, i.e. for all h we set $n_{1,h} = 0.25$. As for the initial prices, we experimented with two settings and obtained very similar results in both of them. One possibility is to choose the same initial prices in all three simulations. In this case, the difference in aggregate outcomes can be attributed only to the difference in the learning parameters. Another approach, which is presented in this paper, is to use the actual prices from the experiment for initialization. Consequently initial prices reported in Table 4.1 are computed as averages of the prices in the first two periods for the corresponding groups.

The parameters of learning over the population of heuristics, and in particular the value of the intensity of choice, are important in determining which pattern is more likely to emerge. In the case of monotonic convergence (top panels), the intensity of choice is low, and the difference in the heuristic performances only slightly affects the evolution of fractions. Thus, the four fractions (and the individual forecasts) remain relatively close together during the simulation causing slow (almost) monotonic convergence of the price to the fundamental equilibrium 60.

[2] The simulation program for the model described in this paper together with brief documentation and configuration settings for different simulations is freely available at `http://www.cafed.eu/evexex`.

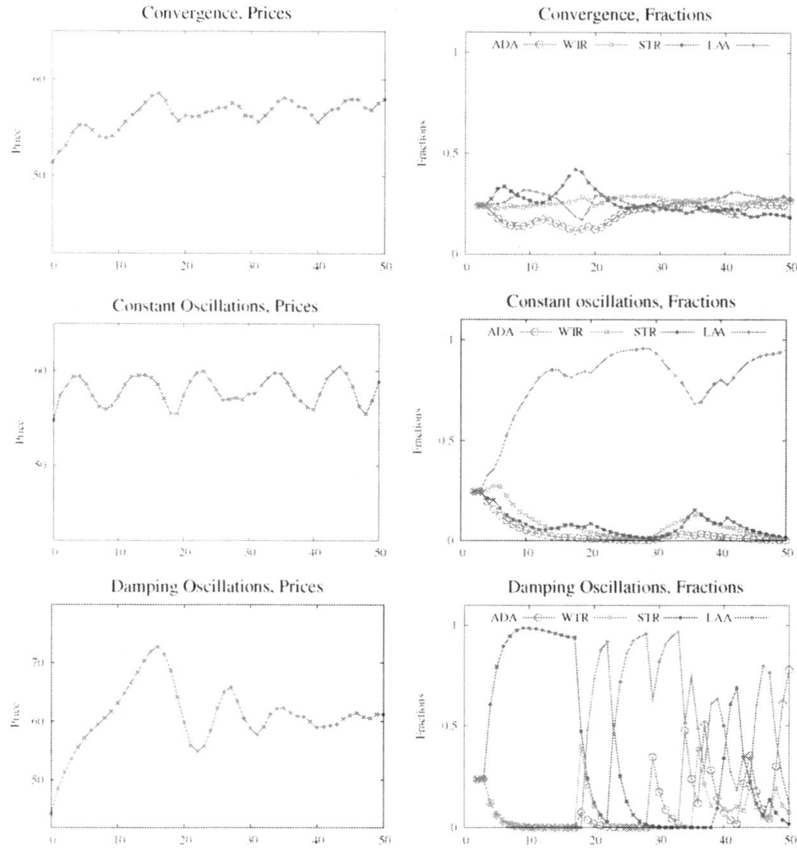

Fig. 4.4 Prices (*left panels*) and fractions of four forecasting heuristics (*right panels*) in the evolutionary model with heterogeneous expectations

Table 4.1 Initial conditions for simulation of different qualitative scenarios reported in Fig. 4.4

	Initial Prices		Heuristics' Initial Impacts				Parameters		
	p_0	p_1	ADA	WTR	STR	LAA	β	η	δ
Convergence	51.36	52.41	0.25	0.25	0.25	0.25	0.3	0.7	0.8
Constant oscillations	54.795	57.415	0.25	0.25	0.25	0.25	1	0.7	0.8
Damping Oscillations	44.265	48.52	0.25	0.25	0.25	0.25	5	0.14	0.5

The increase in price causes a temporary domination of the dynamics by the STR heuristic between periods 14 and 20. However, this rule overestimates the price trend so that, ultimately, three other heuristics take the lead, and price converges to fundamental.

When the intensity of choice increases, a high performance of any heuristic will re-enforce the use of this heuristic. In the second simulation (middle panels) the

LAA heuristic performs very well in the first periods and takes a lead. Moreover, this flexible rule generates a price outcome on which it performs better than other heuristics. For instance, static trend-following rules miss the trend turning points, as it happened around period 36. In this simulation, the fraction of the LAA heuristic dominates the market until the end, explaining coordination of individual forecasts as well as persistent price oscillations around the long run equilibrium level.

In the third simulation (bottom panels), an initial price trend is such that the STR rule dominates and leads to a strong rise of market prices in the first 17 periods. However, ultimately this rule will overestimate the trend, while the more flexible LAA heuristic will predict better. It causes price oscillations in the middle part of simulation with self-confirmed dominance of the LAA heuristic. However, when the intensity of choice is high and the relative weight of the past performances is low, other heuristics may not be completely eliminated. Indeed, to the end of the simulation, the fraction of the LAA heuristic decreases, other heuristics including the ADA rule gain higher weights and the price dynamics stabilizes.

4.5 Conclusion

The simulations reported in this paper illustrate how the interaction and evolutionary selection of individual forecasting heuristics may lead to coordination of individual behavior upon different aggregate market outcomes. The model we presented provides an explanation of very different aggregate outcomes obtained in the same economic setting through the difference in the structural parameters of learning process.

Acknowledgements This work was supported by the ComplexMarkets E.U. STREP project 516446 under FP6-2003-NEST-PATH-1.

References

W. A. Brock and C. H. Hommes. A rational route to randomness. *Econometrica*, 65(5):1059–1095, 1997.

W. A. Brock and C. H. Hommes. Heterogeneous beliefs and routes to chaos in a simple asset pricing model. *Journal of Economic Dynamics and Control*, 22:1235–1274, 1998.

J. Duffy. Experimental macroeconomics. In S. Durlauf and L. Blume, editors, *New Palgrave Dictionary of Economics*. Palgrave Macmillan, New York, 2008. forthcoming.

E.F. Fama. Efficient capital markets: a review of theory and empirical work. *Journal of Finance*, 25:383–417, 1970.

C. Hommes. Heterogeneous agent models in economics and finance. In K. Judd and L. Tesfatsion, editors, *Handbook of Computational Economics Vol. 2: Agent-Based Computational Economics*. Elsevier, North-Holland, 2006.

C. Hommes, J. Sonnemans, J. Tuinstra, and H. v. d. Velden. Coordination of expectations in asset pricing experiments. *Review of Financial Studies*, 18(3):955–980, 2005.

D. Kahneman. Maps of bounded rationality: Psychology for behavioral economics. *American Economic Review*, 93:1449–1475, 2003.

R. E. Lucas and E. C. Prescott. Investment under uncertainty. *Econometrica*, 39(5):659–681, 1971.

J. F. Muth. Rational expectations and the theory of price movements. *Econometrica*, 29(3):315–335, 1961.

T. J. Sargent. *Bounded Rationality in Macroeconomics*. Oxford University Press, 1993.

H. A. Simon. *Models of Man: Social and Rational*. John Wiley, New York, 1957.

V. L. Smith. An experimental study of competitive market behavior. *Journal of Political Economy*, 70(2):111–137, 1962.

V. L. Smith, G. L. Suchanek, and A. W. Williams. Bubbles, crashes, and endogenous expectations in experimental spot asset markets. *Econometrica*, 56(5):1119–1151, 1988.

A. Tversky and D. Kahneman. Judgement under uncertainty: Heuristics and biases. *Science*, 185: 1124–1130, 1974.

Chapter 5
Prospect Theory Behavioral Assumptions in an Artificial Financial Economy

Marco Raberto, Andrea Teglio, and Silvano Cincotti

Abstract We present a model of an artificial financial economy, where a number of heterogenous agents, i.e., households, firms, and a commercial bank make endogenous financial decisions which involve portfolio investments for households, capital structure and dividends policy for firms, and lending and borrowing rates for the commercial bank. Labour income for households and earnings for firms are exogenous determined, according to two independent stochastic processes. Economic policy is set by a government which collects taxes and issues government bonds, and by a central bank which fixes the base interest rate. The purpose of the computational experiments presented in this paper is to focus the attention on a particular and very important aspect concerning the way households form their preferences. In order to include psychological traits in household decision making, Prospect Theory have inspired our study. Indeed, Prospect Theory is a well established framework with a rich theoretical and experimental literature. Computational experiments point out that prospect theory psychological assumptions influence many important financial processes, like assets prices formation and portfolio selection. In particular, loss aversion influences the financial market in a not trivial way. The main results of our study is the emergence of a clear relation between the evaluation period of the household and its portfolio composition in terms of the ratio between risky assets, i.e., stocks, and less risky assets, i.e., bonds. Finally, the model offers more elements to interpret the equity premium puzzle, clearly showing that the stocks-bonds ratio does not depend only from risk-aversion but also from households evaluation periods.

M. Raberto, A. Teglio, and S. Cincotti
DIBE-CINEF, University of Genova, Via Opera Pia 11a, 16145 Genova, Italy, e-mail: [raberto,teglio,cincotti]@dibe.unige.it

5.1 Introduction

This paper presents an agent-based model of a financial economy populated by different types of agents, i.e., households, firms, a commercial bank, a central bank and a government, which interact through a multi-asset financial market. The distinctive feature of the study is the endogenous modelling of financial decisions of agents, who make choices, e.g., portfolio allocation for households or dividend payment for firms, subject only to two exogenous stochastic processes (labour wages for households and returns on investments for firms) and to policy decisions issued by the Government and the Central Bank about taxation and interest rates.

A particular attention is devoted to the balance sheets, considering the dynamics of the financial flows among agents. Firms and bank's equity are divided into shares among households and traded in the financial market. Firms also recur to debt financing, asking for bank loans. The bank collects households deposit and accesses to the standing facilities of the central bank, that sets the interest rate. The government collect taxes and pays bonds coupons to bondholders.

The study presented in this paper focuses on modelling households beliefs formation and households financial preferences. The belief formation process on asset returns takes into account expected cash flows, establishing an endogenous integration between the financial side and the real side of the economy. The model of financial preferences applies some general concepts from *Prospect Theory* (Kahneman and Tversky, 1979; Tversky and Kahneman, 1992), such as the endowment effect (i.e., agents derive utility not from wealth, but from gains and losses defined to some reference level), and loss aversion (i.e., a loss hurts more than an equally large gain produces joy).

Barberis et al. (2001) recently proposed a model of agents financial decision making, according to *Prospect Theory's* psychological assumptions, following the standard consumption-based equilibrium asset pricing framework. In this respect, our approach is different and consists into the integration of a preference structure based on *Prospect Theory* in an agent-based model of a financial economy, in the lines of recent studies in the field, see, e.g., Raberto et al. (forthcoming). In particular, our work is based on the concepts of myopic loss aversion and mental accounting (Benartzi and Thaler, 1995). The main result shown in this paper is the dependence of households portfolio from evaluation periods (i.e., the length of time over which an agent evaluates his portfolio). We show that the stocks prices rise when incrementing the evaluation period, thus confirming the hypothesis that a short evaluation period can be one of the factors that gives rise to the discrepancy between the returns on stocks and fixed income securities. The fact that the combination of a high equity premium, a low risk-free rate, and smooth consumption is difficult to explain with plausible levels of investor risk aversion, has been called by Mehra and Prescott *the equity premium puzzle*, see Mehra and Prescott (1985) for further details. The results we propose corroborate the idea of Benartzi and Thaler that the puzzle could be explained by traders' *myopic loss aversion*, i.e., a combination of loss aversion with a short evaluation period of households (Benartzi and Thaler, 1995).

5.2 The Model

The time structure of the model is the following: two nested time units are considered, let say the day and the month. The month is indexed by τ and is the time span regarding the dynamics of exogenous variables, i.e., the payment of wages and the evaluation of earnings; firms, the commercial bank, the Government and the central bank make decisions on a monthly basis. Conversely, the financial market operates daily and the day, indexed by t, is the time unit considered by households for their financial investments. Each month is supposed to be subdivided into a given number of days (Fig. 5.1).

Firms

Each firm j is characterized by a variable endowment of physical capital A^j_τ and by a time-varying return on physical capital ξ^j_τ; the return on capital is modeled according to an exogenous autoregressive log-normal stochastic process. The product $\xi^j A^j$ gives firm's gross earnings before taxes and interest payment. Physical capital is acquired by means of both equity capital E^j and debt financing D^j; A^j is measured in terms of the same monetary numeraire of E^j and D^j, and its initial amount A^j_0 is determined by the initial level of equity and debt, i.e., $A^j_0 = E^j_0 + D^j_0$.

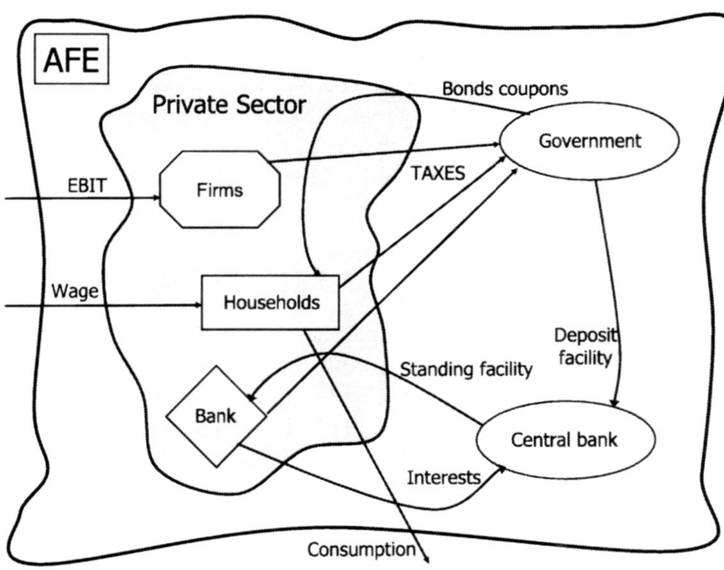

Fig. 5.1 General scheme of the main interactions in the model of an Artificial Financial Economy (AFE). Exogenous flows, along with flows between the private sector and the external institutions are represented

The equity is divided into shares among households and traded in the financial market, the debt is a loan provided by the commercial bank. Net earnings π^j are given by

$$\pi^j_\tau = \xi^j_\tau A^j_{\tau-1} - r^L_{\tau-1} D^j_{\tau-1} - T^j_\tau \tag{5.1}$$

where T^j_τ are taxes paid to the Government on gross earnings, after deducing interest payment, and r^L is the commercial bank lending rate. Net earnings can be paid to shareholders by means of dividends d^j_τ or partially retained to finance new investments in physical capital I^j_τ. New investments can be financed also by bank loans and the issue of new shares; investments in physical capital are made considering the difference between ξ^j_τ and the average cost of capital in the financial market. Generally speaking, firms' financial decision making follows empirically observed managerial behavioural rules, see, e.g., Tirole (2006). Firms are never rationed in the credit market.

Let us note retained earnings with $\widehat{\pi}^j$, i.e., $\widehat{\pi}^j_\tau = \pi^j_\tau - \mathcal{N}^j_\tau d^j_\tau$, where $\mathcal{N}^j_\tau$ is the number of aggregate outstanding shares and d^j_τ is the per share dividend. The dynamics of firms assets and liabilities is thus given by:

$$\begin{cases} A^j_\tau = A^j_{\tau-1} + I^j_\tau \\ D^j_\tau = D^j_{\tau-1} + I^j_\tau - \widehat{\pi}^j_\tau \\ E^j_\tau = A^j_\tau - D^j_\tau . \end{cases} \tag{5.2}$$

The Commercial Bank

The commercial bank collects households deposits B_τ, provides loans L_τ to firms, and holds a buffer account C_τ at the central bank, which can be positive or negative. The commercial bank sets the lending rate r^L to firms according to a mark-up rule on the central bank policy rate r, i.e., $r^L = \mu_L r$, where $\mu_L > 1$ is the mark-up. The rate on households deposits r^B is determined by $r^B = \mu_B r$ where μ_B is lesser than one. Net earnings are given by

$$\pi^b_\tau = r_{\tau-1} C_{\tau-1} + r^L_{\tau-1} L_{\tau-1} - r^B_{\tau-1} B_{\tau-1} - T^j_\tau \tag{5.3}$$

where T^j_τ are taxes as a fraction of gross earnings paid to the Government. The capital structure of the bank is composed by both equity capital E^b and debt financing, i.e., the Central Bank account and households deposits. The bank equity is divided into shares among households and traded in the financial market. Given the amount of L and B set by firms and households, respectively, and the dynamics of equity $E^b_\tau = E^b_{\tau-1} + \widehat{\pi}^b_\tau$, where $\widehat{\pi}^b$ are the retained earnings, the bank adjusts C according to the budget constraint $C_\tau = E^b_\tau + B_\tau - L_\tau$.

The Government and the Central Bank

The Government runs a financial budget: income is given by taxation of firms and commercial bank earnings; expenditures depend on unemployment benefits and on the interest rates on government debt. The government may issue both short-term or long-term bonds in order to finance the budget deficit. Bonds have a face value which is paid at the maturity date, and pay fixed coupons to bondsholders anchored to the central bank policy rate. The central bank implements monetary policy decisions by means of a policy rate r which is used both as a borrowing or lending rate for the commercial bank account. The goal of both the Government and the central bank policies is the pursuit of low volatility in the asset market and of long-run growth in the economy by means of accumulation of physical capital by firms.

Households

Households are simultaneously taking the roles of workers, consumers and market traders. They receive an exogenous labor income at a common wage if employed and an unemployment subsidy if unemployed, and determine how much to spend and how much to save. Households' saving-consumption decisions are modelled according to the theory of buffer-stock saving behaviour (Deaton, 1992), which states that households consumption depends on a precautionary saving motive, determined by a target level of wealth to income ratio. Households can either invest their savings in the asset market, by trading stocks or bonds, or can put them in a saving account that pays a fixed, risk-free interest rate. Households form beliefs about assets returns considering a common forward horizon of three months. Besides, each household i is characterized by an evaluation period ε_i which is a multiple of the forward horizon and is used to compute preferences and evaluate investments (Benartzi and Thaler, 1995). Beliefs are formed according to three stylized behavior, i.e., random, chartist and fundamental. In particular, expected asset returns for each asset j, issued by the j-th firm, are given by a linear combination of three terms: a scalar random component $\rho^r_{j,i}$, a set of past returns $\rho^c_{j,i}$ computed in a backward time window, and a fundamentalist scalar term $\rho^f_{j,i}$. In order to compute the fundamental return, each household estimates a fundamental price

$$p_{j,i} = (E^j_\tau + \widehat{\pi}^j)/\mathcal{N}^j \tag{5.4}$$

taking into account the equity capital of firm j and the expected retained earnings in the forward horizon. Given the fundamental price and considering the last market price, the household derives the expected fundamental return $\rho^f_{j,i}$. Composing the three terms and adding expected cash flow yields $y^e_{j,i}$ (i.e., dividends for stocks and coupons for bonds), households determines a set of total expected returns $\rho_{j,i}$ as

$$\rho_{j,i} = \alpha^r_i \rho^r_{j,i} + \alpha^c_i \rho^c_{j,i} + \alpha^f_i \rho^f_{j,i} + y^e_{j,i} \tag{5.5}$$

where α_i^r, α_i^c and α_i^f are household's weights that sum to one. Then households build a normalized histogram $H[\rho_{j,i}]$ where the set of total expected returns is grouped in M_i bins. It is worth noting that a large number of bins M_i means that the household is more careful when examining the asset's past performance, taking into account more elements (it uses a higher resolution to build the histogram).

The histogram $H[\rho_{j,i}]$ can be seen as a prospect $\mathscr{P} = [\rho_{j,i}^H, p_{j,i}^H]$ where $\rho_{j,i}^H$ are the bins center values of the expected total returns histogram and $p_{j,i}^H$ are the associated probabilities, i.e., the level of the normalized histogram. If the evaluation period of the household is longer than the forward horizon used in the beliefs formation, it means that the prospect should be iterated accordingly. To this aim, we modelled how the structure of a prospect varies when the evaluation period changes. Following the concepts of myopic loss aversion, we introduce a new prospect $\mathscr{P}^n$ that represents the mental accounting (see Benartzi and Thaler, 1995) of the agent when considering the risky investment, that means an n times iteration of prospect $\mathscr{P}$. Accordingly, the number of elements of the iterated prospect $\mathscr{P}^n$ will pass from M_i to $\mathscr{M}_i$. Thus, each household will face a new prospect $\mathscr{P}^n = [\rho_{j,i}^{H_n}, p_{j,i}^{H_n}]$ depending on its evaluation period.

Prospect theory utility is defined over gains and losses, i.e., returns ρ^{H_n}, rather than levels of wealth. The value function for the ith household has the following form:

$$v_i(\rho_{j,i}^{H_n}) = \begin{cases} (\rho_{j,i}^{H_n})^\alpha & \text{if} \quad \rho_{j,i}^{H_n} \geq 0, \\ -\lambda_i(-\rho_{j,i}^{H_n})^\beta & \text{if} \quad \rho_{j,i}^{H_n} < 0, \end{cases} \tag{5.6}$$

where λ_i is the coefficient of loss aversion of household i. By means of behavioral experiments Kahneman and Tversky (1979) estimated α and β to be equal to 0.88 and λ to be equal to 2.25.

Given the histogram of composed expected returns, the ith household may calculate the utility of asset j as,

$$U_{j,i} = \sum_{\mathscr{M}_i} p_{j,i}^{H_n}\, v(\rho_{j,i}^{H_n}), \tag{5.7}$$

where $p_{j,i}^{H_n}$ are the probabilities associated to $\rho_{j,i}^{H_n}$. These utilities are finally normalized and mapped into assets weights by means of a linear transformation. Once the assets weights are available, the household can build its desired portfolio and emit orders consequently. Orders are therefore submitted to a clearing house that determines assets new prices.

5.3 Results and Discussion

We present a what-if analysis that considers the variation of the assets price levels with respect to changes both in traders psychology and in policy strategies. The attention will be not only on the general price level of the assets in the financial market, but also on the relative prices between different assets with particular reference to the relation between firms stocks and government bonds. We focus our analysis on three main parameters of the model.

The first one, which is purely psychological, is the loss aversion value λ. As described in Sect. 5.2, the loss aversion is a crucial aspect of the approach adopted from prospect theory when describing decision under risk conditions. In our model, irrespective of traders beliefs, loss aversion indicates the tendency to be more sensible to reductions in the level of well-being than to increases, when buying risky assets. The first study we present shows how the prices of different assets do vary when agents have different loss aversion.

The second parameter that we consider in our computational experiments is a purely economic one: the central bank policy interest rate r. The interest rate r affects the interest rate on loans r^L and the households borrowing rate r^B, thus having significant effects on the model dynamics, and in particular in households portfolio choices. It is worth noting that changes in both the loss aversion and the central bank interest rate can produce similar consequences in households financial decisions.

The third study we propose regards the variation of a more delicate parameter of the system: the households evaluation period ε_i, which denotes the estimated period of time an household thinks to keep its investments in financial assets. As it will be shown, this variable has an important impact on the dynamic of the financial market, and could give an explanation of the equity premium puzzle (see Benartzi and Thaler, 1995), at least under some of the psychological assumption of prospect theory.

The simulations we present refer to a model populated by 2,000 households and three firms. Five assets are traded in the financial market: three firms stocks, the bank stock, and a long term government bond. Firms are endowed with a constant physical capital and make no new investments. The return on physical capital ξ is modeled by means of an autoregressive process of order 1 with mean 0.06 and memory parameter 0.9. Firms are characterized by different dividends pay-out strategy; firm 1 pays 80% of its net earnings to shareholders, while firm 2 and firm 3 pay 90% and 100%, respectively. Retained earnings are used to increase the equity base of each firm. Among traders, fundamentalists and chartists are 5% each, while the rest are random traders. The commercial bank mark-up μ_L is 1.5, while μ_B is set equal to 0.8. The bank dividend policy is to pay 100% of its net earnings. The government applies a fixed tax rate of 15% both on capital income for households and on corporate earnings of firms and bank. The government bond maturity date is set at the end of the simulation. Finally, each month is considered to be subdivided in five trading days.

Firms and bank balance sheets have been initialized in order to characterize all stocks by the same initial fundamental price, which have been set to 100. In particular, the initial equity for each of the three firms is equal to 10,000,000, while the bank equity has been set to 6,000,000; besides, the number of shares outstanding is 100,000 for each firm, and 60,000 for the commercial bank. Currency units are arbitrary. Each firm is endowed with an initial debt of 20,000,000, so that the

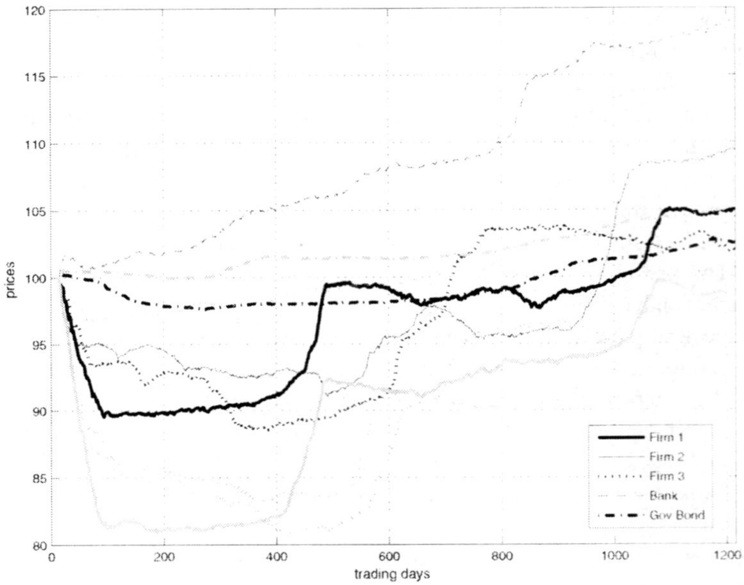

Fig. 5.2 Assets price levels. The reference plots are in black. The grey lines represent price levels when the loss aversion raises from $\lambda = 1.5$ to $\lambda = 2.5$. Currency units are arbitrary

aggregate amount of loans by the commercial bank to firms is 60,000,000, which, given the bank equity, corresponds to a core tier 1 ratio of 0.1. Households are initially endowed with an equal number of shares for each asset and a bank account of 5,000.

Figures 5.2, 5.4 and 5.5 present the assets price levels (black lines) of the reference simulations and price trajectories (grey lines) related to the three alternative simulation scenarios described before. In the reference simulation, the central bank interest rate is $r = 0.05$, loss aversion λ is set to 1.5 for all households, and households evaluation period corresponds to 2 times the forward looking window, i.e., $\varepsilon_i = 6$ months. A clear and relevant evidence of the basic simulation is that the bond price trajectory exhibits a lower volatility with respect to stocks prices. It is worth noting that this is an endogenous emerging feature which has not been modeled in advance, but mainly depends on two factors: first, the face value, which is a strong anchoring to the expectations of price dynamics, and second, the bond cash flow which is characterized by constant coupons, while stocks pay varying dividends whose value is a fixed fraction of the stochastic process of earnings. Furthermore, it emerges that the price trajectories are clearly affected by the cash flows of different assets, see Fig. 5.3. In particular the commercial bank, which pays the highest dividends, generally results to be the best performing stock. Finally, it is worth remarking that the shape of the price processes exhibit jumps, crashing and periods of low volatility, realistic features which clearly depend on the interplay of random, chartist and fundamental strategies.

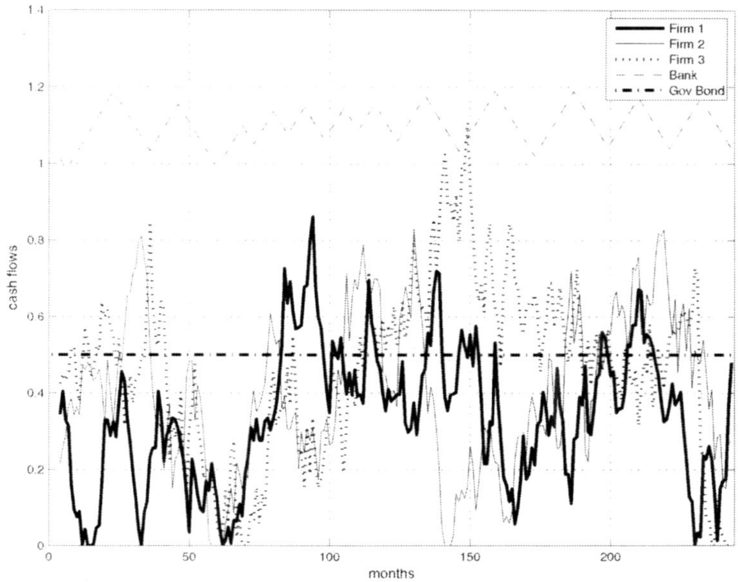

Fig. 5.3 Assets monthly cash flows, i.e., dividends for stocks and fixed coupons for the government bond. Currency units are arbitrary

The grey lines in Fig. 5.2 show the price trajectories referred to a simulation where households loss aversion is raised from 1.5 to 2.5. The loss aversion growth clearly reduces the stocks price level, increasing on the other hand the price of the government bond. The higher volatility of stocks determines such effect, because households tend to overestimate the risk of loosing money with respect to the previous case of $\lambda = 1.5$. It is worth noting that this effects is stronger in the initial phase of the simulation, when households feel that the stocks price is too high, but later, due to the good returns of stocks with lower prices, stocks prices tend to raise again, getting closer to bonds prices. For lower loss aversions we obviously obtain the opposite effect.

In Fig. 5.4, the reference price trajectories (black lines) are compared to the trajectories (grey lines) referred to a simulation where the central bank adopts a looser monetary policy, setting a lower interest rate of $r = 0.04$. Being the lending rate r^L to firms and the borrowing rate r^B to households dependent on the central bank policy rate r, a change on the policy rate determines strong effects in the stock market. The effect of the new monetary policy by the central bank is twofold. A lower level of r implies less incentives for households to buy government bonds, and to keep savings on their bank account due to lower r_B. Conversely, a lower level of r_L reduces the debt cost for firms, increasing net earnings and dividends, thus making firms stocks more appealing. Finally, the commercial bank faces a reduction of both costs (due to r and r_B) and revenues (due to r_L), but the reduction of revenues overcomes the reduction of costs because of the fixed mark-up, causing a reduction of

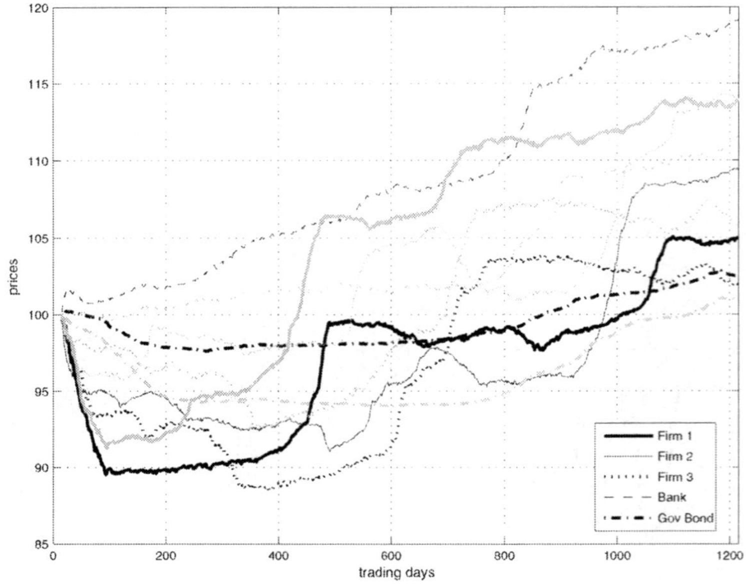

Fig. 5.4 Assets price levels. The reference plots are in black. The grey lines represent price levels when the central bank policy rate decreases from $r = 0.05$ to $r = 0.04$. Currency units are arbitrary

bank earnings. These considerations are clearly reflected in the assets price dynamics of Fig. 5.4 (grey lines), where the higher demand for firms stocks makes the stock prices rise with respect to black lines, while both the bond price and the bank stock remain under the level related to $r = 0.05$.

The case of Fig. 5.5 require a more careful discussion. The reference plots are here superimposed to prices trends that refer to longer households evaluation periods. Their length corresponds to 5 times the forward horizon (grey lines) instead of 2 times as in the reference plots (black lines). Households are therefor less myopic. According to the definition of Benartzi and Thaler, myopic loss aversion is the combination of two concepts from the psychology of decision making: loss aversion and mental accounting. Mental accounting, a concept first named by Richard Thaler (1980), attempts to describe the process whereby people code, categorize and evaluate economic outcomes. In our case, the idea is that the longer the investor aims to hold the asset, the more attractive the risky asset will appear. In other words, two factors contribute to an investor being unwilling to bear the risks associated with holding equities, loss aversion and a short evaluation period. This combination is referred as myopic loss aversion. Figure 5.5 shows that, given both the same beliefs structure and the same loss aversion value, a longer evaluation period determines a growth in the stocks price levels and a reduction in the bond price level. This result is coherent with the approach of Benartzi and Thaler towards the *equity premium puzzle*, confirming that the myopic loss aversion could be a plausible explication.

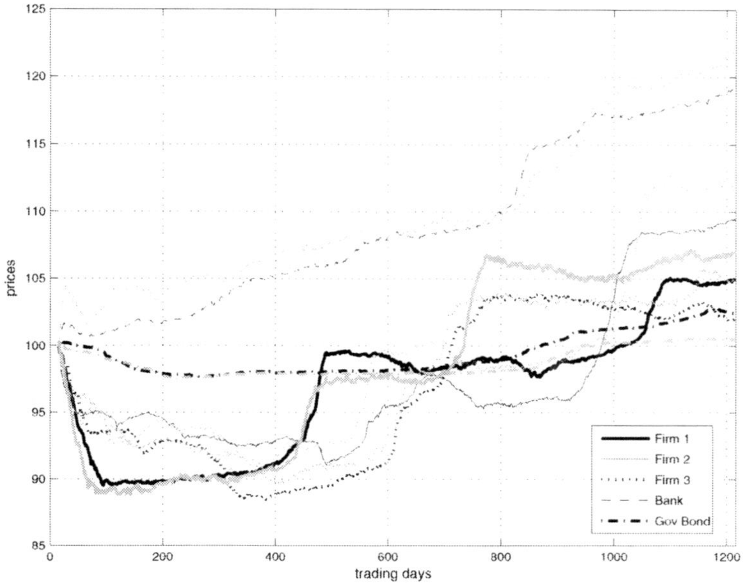

Fig. 5.5 Assets price levels. The reference plots are in black. The grey lines represent price levels when households evaluation period increases by a factor 2.5. Currency units are arbitrary

5.4 Conclusions

In this paper it has been presented a model of an artificial financial economy where the financial decisions of the agents are endogenously taken. The model is particularly complete in terms of agents and in terms of their economic interaction. The agents acting in the system are the government, a central bank, a commercial bank, some firms and many households. The main financial flows among the agents are represented in the model; including standing facilities that depend on the interest rate set by the central bank, loans to the firms from the commercial bank, deposits of household into the bank, bond coupons payments and tax collection by the government, and other interactions that has been described in the paper. The equity of both the commercial bank and the firms is traded in the financial market.

A particular attention has been devoted to the modelling of the beliefs formation mechanism of households trading in the asset market, and on their preferences structure that is designed according to *Prospect Theory*.

The computational experiments have been conceived so to study the reaction of the assets price levels to changes of some parameters of the model.

The experiments clearly show the dependence of the asset prices on the households loss aversion. When households are more loss averse, the price levels of stocks tend to fall, mostly due to a negative perception of the higher volatility of stocks, that induces them to rebalance their portfolio buying government bonds or keeping money in the bank account.

On the other hand, a lowering of the policy interest rate by the central bank affects both the lending rate to the firms and the rate on households deposits. Being the bond coupons linked to the central bank interest rate, households prefer to invest on stocks rather that on government bonds, letting the stock price level rise. For similar reasons, households tend to keep less money in their bank account.

The last what-if analysis we present is intended to give a contribution to the comprehension of the *equity premium puzzle*, articulated more than twenty years ago by Mehra and Prescott. Inspired by an idea of Benartzi and Thaler, we showed that the equity premium increases when extending the evaluation period of households. This means that when households think to hold their assets for a longer period of time, their propensity to invest in more risky assets (stocks) rises. According to our model, the relatively low share of stocks in households portfolio, that seems inconsistent with the high stock returns, can be explained supposing short evaluation periods for households.

Finally, we are confident that our model is a rich and solid framework where performing further scientific experiments in order to get a better understanding of the economic key aspects underlying the dynamics of a financial economy.

Acknowledgements This work has been partially supported by the University of Genoa and by the European Union under IST-FET STREP Project EURACE.

References

N. Barberis, M. Huang, and T. Santos. Prospect theory and asset prices. *Quarterly Journal of Economics*, 116(1):1–53, 2001.

S. Benartzi and R. H. Thaler. Myopic loss aversion and the equity premium puzzle. *The Quarterly Journal of Economics*, 110(1):73–92, 1995.

A. Deaton. Household saving in ldcs: credit markets, insurance and welfare. *The Scandinavian Journal of Economics*, 94(2):253–273, 1992.

D. Kahneman and A. Tversky. Prospect theory: an analysis of decision under risk. *Econometrica*, 47(2):263–292, 1979.

R. Mehra and E. Prescott. The equity premium puzzle. *Journal of Monetary Economics*, XV:145–161, 1985.

M. Raberto, A. Teglio, and S. Cincotti. Integrating real and financial markets in an agent-based economic model: an application to monetary policy design. *Computational Economics*, 2008, available online, DOI: 10.1007/s10614-008-9138-2.

R. H. Thaler. Towards a positive theory of consumer choice. *Journal of Economic Behavior and Organization*, 1:39–60, 1980.

J. Tirole. *The Theory of Corporate Finance*. Princeton University Press, Princeton, 2006.

A. Tversky and D. Kahneman. Advances in prospect theory: cumulative representation of uncertainty. *Journal of Risk and Uncertainty*, 5(4):297–323, 1992.

Chapter 6
Computing the Evolution of Walrasian Behaviour

Gonzalo Fernández-de-Córdoba and Álvaro P. Navas

Abstract The main solution concept in economics is the Walrasian equilibrium. However, Walrasian theory builds upon the central hypothesis of price taking behaviour of the agents of the economy. This hypothesis precludes strategic behaviour of manipulating prices in the agent's own advantage, and it is usually linked to the intuition that in an economy with a large number of atomless agents, their ability of manipulating prices is negligible. Results of convergence to the Walrasian equilibrium with a low number of firms have been much celebrated. In this paper we show that without imitation or mutation, but with sensible behavioural rules, it is possible to design an artificial economy, where a low number of firms can attain the Walrasian allocation in finite time, and much more.

6.1 Introduction

The main solution concept in economics is the Walrasian equilibrium. However, Walrasian theory builds upon the central hypothesis of price taking behavior of the agents of the economy. This hypothesis precludes strategic behaviour of manipulating prices in the agent's own advantage, and it is usually linked to the intuition that in an economy with a large number of atomistic agents, their ability to manipulate prices is negligible.

Vega-Redondo (1997) showed that in an evolutionary context where the behaviour of a finite number of firms, producing an homogeneous product, characterized by (a) imitation of the most successful observed strategy, and (b) experimentation,

G. Fernández-de-Córdoba
Departamento de Economía e Historia Económica, Universidad de Salamanca, Edificio F.E.S. Campus Miguel de Unamuno, E-37007 Salamanca, Spain, e-mail: gfdc@usal.es

Á. P. Navas
UAV Navigation SL. Av. Severo Ochoa no45, Alcobendas, Madrid, 28100, Spain, e-mail: anavas@uavnavigation.com

would, in the long run, yield the Walrasian outcome even with small populations, that is, populations where particular individuals are a significant fraction of the population.

The importance of V–R's contribution is two fold: as mentioned above, his result shows that in an evolutionary context the atomless heuristics can be removed, but also and more importantly, the requirement on rationality of the agents is reduced as it is common in evolutionary literature. Performance, imitation and mutation are all what is needed to achieve the Walrasian behaviour.

However, as implied by Chatoe (1998), the mapping between theory and evolutionary simulations, should allow a description of rationality that is genuinely bounded and procedural. In this paper we modify the basic V–R's artificial economy, to compare the performance of a given set of firms endowed with different procedural rules. As in Vega-Redondo (1997) it is natural to think of firms comparing relative performance as a way of taking decisions, however, it is not the only one we could consider. For example, it would also be natural to think of firms comparing average returns, minimum returns, positive returns, or any other measure that requires the same amount of information, i.e., to allow firms to filter information and behave correspondingly in many other different ways than simply comparing to the best performing firm of the past round. The behavioural rules explored in this paper are given to firms and do not compete among them as in Chavalarias (2006).

This paper explores how different behavioural rules select strategies in different economic environments where output levels are identified with the elements of the strategy set of firms. Simulations show that under some circumstances the unique surviving strategy is the Walrasian output, and this occurs in finite time (as opposed in V–R's economy) regardless of the number of firms.

However, the idea that the Walrasian equilibrium can be rescued by means of an evolutionary algorithm is of a misleading generality. This is so because the environment where the evolutionary techniques apply is of much more generality than the one presented by V–R to sustain the Walrasian equilibrium. Once this generality is introduced in a very similar economy, using different behavioural rules, the Walrasian allocation appears with a different meaning than that given by the orthodox explanation. Furthermore, our application shows that, in contexts where the Walrasian equilibrium would yield negative profits, those behavioural rules that would have selected the Walrasian equilibrium provide a meaning to the solution found.

The first part of this paper, Sect. 6.2, is devoted to explain the basic set-up of Vega–Redondo's economy, and to provide a computational tool for evolutionary simulation and experimentation with an economy of the type described by Vega–Redondo. Such a tool can prove useful given the analytical difficulties to link a specific choice of parameters (number of firms, specification of market demands, costs structures, etc.) with results such as the speed of convergence, or the fraction of time in which those states that are stochastically stable should appear.

The second part of this paper, Sect. 6.3, presents the results of several simulation experiment of an economy where agents may have different behavioural rules. In our economy agents compare the profits obtained in each iteration to those obtained by all firms in that iteration. Those firms that obtain a relatively low profit eliminate the strategy used from their respective strategy set. In the next round they mutate by selecting a strategy from those remaining in their strategy set. Since in each iteration at least one strategy is eliminated, and there exist a finite number of strategies, our simulation experiments show that the process of elimination of strategies produces a stable selection in the strategy set, and this selection occurs in finite time. Notice that elimination of strategies as a selection device substitutes imitation of the best performing firm. We accommodate the large evidence on convergence of oligopolistic markets where no imitation seems to take place in the observed behaviour of agents as illustrated by Boshc-Domenech and Vriend (2004), where they provide damaging evidence against the prevalence of imitation. Section 6.4, sketches a formal argument already explored in Fernández-de Córdoba showing that no imitation is required to obtain the Walrasian allocation in an evolutionary context.

6.2 The Vega–Redondo Economy Model

The description of the model follows closely that of Vega-Redondo (1997). Consider a set of N firms indexed by $i = \{1, 2, \ldots, N\}$ competing in a market where an homogeneous commodity Q is sold. The demand is summarized by an inverse demand function that we shall assume to be linear and decreasing $P = a - bQ$. For every total output supplied to the market, this demand function provides the market clearing price $P(Q(t))$ at which it is sold. All firms are assumed to be ex-ante symmetric and with identical cost function $C(q)$, that we shall assume to be an instance of the class of functions $C(q) = c_1 q(i)^{c_2}$, where $q(i)$ is the individual production of firm $i = \{1, 2, \ldots, N\}$, and the parameters c_1 and c_2 are non-negative. It is assumed that firms choose their output from a common finite grid called strategy set $S = \{0, s, 2s, \ldots, Ks\}$; where $s > 0$ and the pair (K, s) is specified so that the existing and unique symmetric Walrasian individual output belongs to the grid.

The evolutionary dynamics proceed in discrete time, which is indexed by $t = 0, 1, 2, \ldots$ The current output profile induces a profit profile obtained as follows: $P(Q(t))q_i - C(q_i)$, $i = \{1, 2, \ldots, N\}$. Once profits are realized, firms either imitate the strategy of the firm who has attained the maximum profit or alternatively mutate with some common independent probability $m(t) > 0$.

Summarizing the information of the elements of the theory in the standardized form (see Smith, 1989, for a justification): the environment is given by a cost function as described above for each $i = \{1, 2, \ldots, N\}$ firm, and a given aggregate demand function. The institution is defined by a market structure where firms make bids in quantities, i.e., their message is the amount they are willing to supply in each period, and the price at which transactions take place is the result of the projection into the demand function of the aggregated supply. Firms are, therefore, obliged to sell the announced quantity at the resulting demand price. The behaviour of agents is described by the selection of a single strategy available in their strategy set at each time period. This selection of strategy is random over the elements of strategy set. In the first period, and at each period the chosen strategy is put under a test: if the selected strategy provides the highest profit, this strategy is imitated by all firms (including the firm that selected it) with probability $1 - m(t)$, and with probability $m(t)$, firms mutate and choose randomly from the common strategy set. The information requirement is that all firms have knowledge about the common strategy set, as well as who did what and how much profit each firm earned, in order to be able to imitate the most successful strategy. Vega-Redondo (1997) demonstrates that such economy will stabilize in the long run around a monomorphic state where all firms supply the Walrasian output.

6.3 The Behavioural Rules Set

In this section we depart from V–R's artificial economy. Through a series of simulations we explore the consequences on the stable strategy set when different selection devices of strategies replace imitation in the behaviour of firms. Firms, rather than imitating the best performing firm, in each round delete those strategies that do not pass a given filtering criteria. The environment and the market institution are as in the V–R economy, however there is a modification in the behaviour of the agents. In the following sections the behaviour is changed to accommodate different behavioural rules, and serves to replace imitation and mutation by strategy deletion. In this behaviourally modified economy, in the first round firms randomly choose a strategy from a common strategy set. The strategy chosen is then put under a test: if the chosen strategy performs well according to the exogenously given criteria (the filtering criteria), then the strategy survives that round, otherwise the strategy is deleted from the strategy set, and a new strategy is selected, from the idiosyncratic strategy set in the next round.

The developed interface (available from the authors), takes as inputs two sets of parameters: (a) environment parameters, and (b) grid thickness and number of iterations.

The filters for which we provide interface are the following:

(a) Average Profit Filter: In the first round firms choose randomly from a common strategy set. This set contains as elements a finite grid that includes the Walrasian allocation. The first round strategy profile induces a unique market-clearing price, and given the cost parameters, a profits profile is induced. Those firms whose profits do not reach the average profit delete the strategy chosen. After the first round, strategy sets are firm-specific. In the second and subsequent rounds, firms choose randomly from their strategy sets, filtering again with the average profit filter.

Simulations show that the unique surviving strategy is the Walrasian output, and this occurs in finite time regardless of the number of firms, demand parameters, and costs parameters in the absence of fixed costs. This is a result of the paper and deserves some attention.

Notice that the behavioural elements differ in some crucial aspects with respect to the Vega–Redondo's economy: first, selection comes from deletion of strategies rather than from imitation, second, the mutation rate is one for every firm at every period, and third, there is no need to have information about the strategy that gave certain profits to a specific firm, just average profit has to be known by firms, and fourth, convergence to the Walrasian allocation occurs in finite time.

The average filter provides a relative measure of performance and, as we will see, relative measure is what drives the results of this paper. To fix some intuition about our result in comparison to V–R's lets assume that all firms are exactly on the Walrasian allocation, any individual deviation above the Walrasian allocation decreases the sale price, while it increases the costs of the deviating firm. All firms might go into losses (with linear costs this is certainly the case), but the firm that deviated will incur in the lowest profit of that round. In V–R's economy, this deviating firm will imitate any other firm, and in our economy the strategy chosen by the deviant firm will be deleted because it pays off less than the average profit. If on the contrary the deviation undercuts production, there will be an increase in prices, and those firms producing more will have higher profits, triggering imitation by the deviant firm in V–R's economy and deletion in ours.

As an example we show next the result of a run with following parameter values.

Demand is linear with 45° slope, and costs are quadratic. The number of firms is six, and the grid size is chosen so that the initial common strategy set has 51 elements, with individual Warasian output $q_j^w = 12.5$, lying in the middle of the set, with 25 elements below that output and 25 elements above. Profits in the Walrasian allocation are 156.25 (see Table 6.1).

We see in Fig. 6.1, the histogram provided by the interface. The x-axis contains the strategy set, and the y-axis is the frequency of each surviving strategy at the end of the run, or equivalently, the number of firms that are playing a strategy in S. As we can see, the frequency of the Walrasian output is equal to the number of firms N, that is, the Walrasian output is the unique strategy that survives the deletion process.

(b) Minimum Profit Filter: This filter is less selective than the average since, in each round, only those firms obtaining the minimum profit delete that strategy. As an example of the output in the histogram window of the interface we choose the following parameters (see Table 6.2).

Table 6.1 Parameter values

a	b	c_1	c_2	N	Grid size
100	1	1	2	6	51

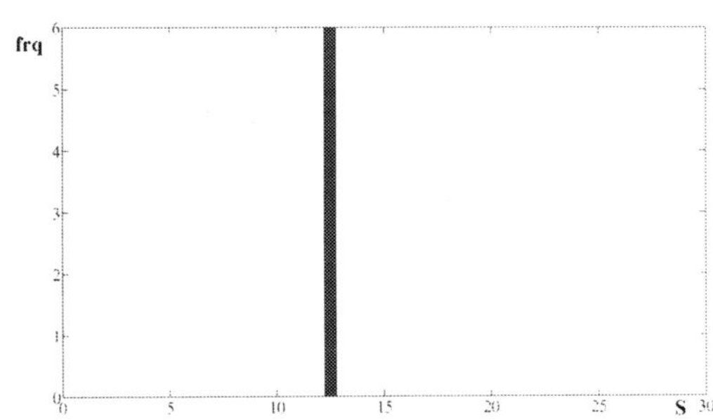

Fig. 6.1 Unique surviving strategy for the six firms

Table 6.2 Parameter values with 3 firms

a	b	c_1	c_2	N	Grid size
100	1	1	1	3	51

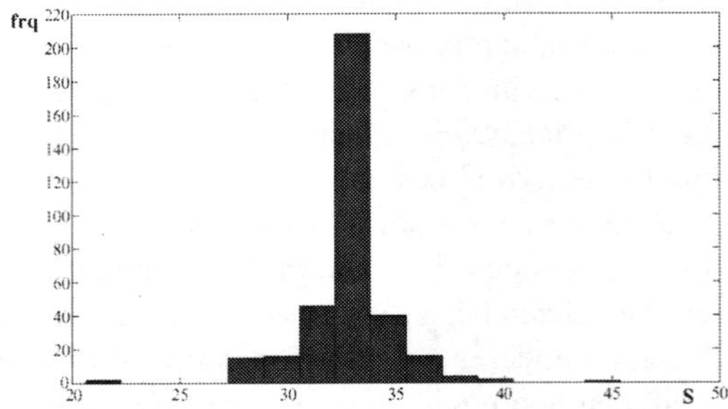

Fig. 6.2 Surviving strategies for 3 firms in 100 runs

Table 6.3 Parameter values used in the Walrasian profit filter

a	b	c_1	c_2	N	Grid size
100	1	1	1	3	51

The values for the demand parameters provide a simple picture of a demand function with 45 degrees negative slope and total market size of 100: Costs parameters imply constant marginal costs and the number of firms is 3. These values imply a Walrasian allocation $q_j^w = 33$.

Figure 6.2 shows 100 repetitions of the experiment and the frequencies of the surviving strategies. The x-axis is the strategy set, and the y-axis contains the number of times that a strategy appeared as an element of the surviving strategy set of a firm after the run. As we can see in Fig. 6.2, most of the time the Walrasian allocation is contained in any of the three strategy sets, but sometimes it is either not the unique value, or the singleton strategy set for a firm is different than the Walrasian output.

(c) Walrasian Profit Filter: For any given parameter values, the Walrasian profit is computed. Firms delete from their strategy sets those strategies that, in a given iteration, do not reach that level. The Histogram window of the interface presented above would produce a typical single run like in the next figure. The parameters chosen are shown in Table 6.3.

Clearly, if most firms produce an output level below the Walrasian output, the profits for these firms will be higher than the Walrasian profit, and therefore those strategies are not deleted. However, this pattern is more clearly seen after 100 runs. The next figure shows that most of the time, the surviving strategies are below the Walrasian output.

Figure 6.3, illustrates a common feature in evolutionary models, namely, the existing difference between absolute versus relative performance. The Average Profit Filter, a relative-to-the mean rule, selects the Walrasian allocation as the unique outcome, inducing Walrasian profits, however, the absolute performance rule

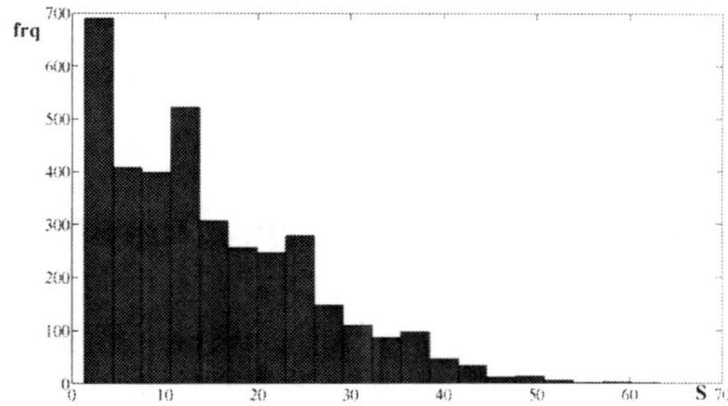

Fig. 6.3 Surviving strategies in 100 runs with the Walrasian profit filter

Table 6.4 Parameter values with a zero profit filter

a	b	c_1	c_2	N	Grid size
100	1	1	100	3	51

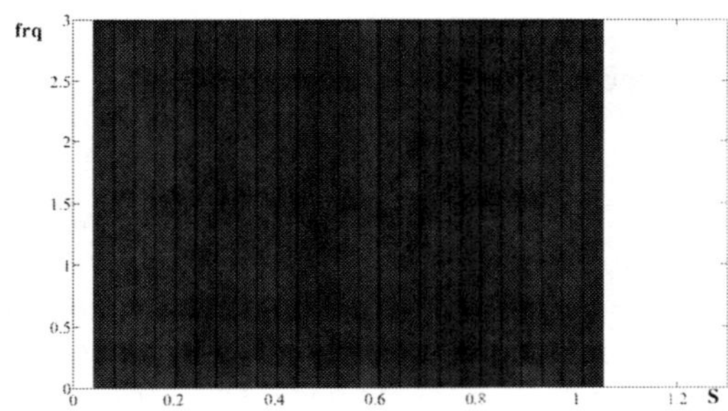

Fig. 6.4 Surviving strategies with the zero profit filter

given by the Walrasian Profit Filter, not only does not select the Walrasian allocation, but also imposes a penalty on it.

If real subjects were placed in an experiment with the environment and institutional framework of this section, and if they were using an absolute measure of performance, like the Walrasian profits, then we would observe a strategy profile closer to a collusive profile rather than a competitive one.

(d) Your Choice Filter: Fixes the minimum acceptable profit. The default value in the interface is zero, meaning that firms are willing to be active in the industry if they have no losses. The parameter values as shown in Table 6.4 induce a Walrasian output $q_j^w = 1$.

We see in Fig. 6.4, that under the zero profit behavioural rule, any output below Walrasian output is left in the strategy set of all firms, with the exception of zero output. At the right of the Walrasian output there is a single strategy. However, with even larger marginal costs, this strategy is also filtered out.

Notice, that under this behavioural rule, profits are larger than those attained in the Walrasian allocation, and that this filter is the less demanding in terms of the information required to its' implementation. Firms just need to know their actual profits and the desired minimum profits level. Notice also that under linear costs this choice coincides with the Walrasian Profit Filter, since zero profit is the implied profit level under Walrasian output with constant marginal costs.

6.4 Walrasian Equilibrium Revisited

If the behavioural rule of the average filter we have assumed in our simulations yields as a result a value that is consistent with the behaviour we would have obtained if agents were maximizing relative profits, then the Walrasian equilibrium should be obtained also as the result of this behaviour. In fact it does.
The individual profit function can be written, in general, as:

$$\Pi_i = P\left(\sum_k q_k\right) q_i - C(q_i) \tag{6.1}$$

And the relative (to the mean) profit as:

$$d_i = P\left(\sum_k q_k\right) q_i - C(q_i) - \frac{1}{N}\sum_j \left[P\left(\sum_k q_k\right) q_j - C(q_j)\right] \tag{6.2}$$

Since there are no externalities in production, the effect on prices of one firm changing its output is completely offset by another firm, and the resulting equation, after maximization and assuming a symmetric equilibrium, is simply: $P = C'(q_i)$, that is, price equal to marginal costs as in the Walrasian allocation.

We see that the Walrasian equilibrium can be obtained 'as if' agents were maximizing relative profits with no imitation. As can be seen from the above equations, with identical firms the relative to the mean measure is equivalent to the absolute difference in profits between any two firms, and as a result agents imitating the most successful firm of the past round behave like agents deleting those strategies that do not perform as well as the mean firm.

6.5 Conclusions

In this paper we have studied a very similar artificial economy to that proposed by Vega-Redondo (1997). We have removed the assumption of imitation with the use of different behavioural rules. These rules have in common that those selected strategies that do not perform well accordingly to given criteria are filtered out from the firm-specific strategy set. We have obtained a number of interesting results as an outcome. The first one is that using the average filter we obtain the same result of Walrasian selection in finite time with no imitation. This result is, however, very natural: Imagine a firm that is playing the Walrasian strategy, and all other firms are producing 'too much'. In this context prices will be low, potentially inducing negative profits for all firms. Since the firm producing the Walrasian output level is producing less, its profits would be relatively higher. Imagine, on the other hand, that all other firms are producing below than the Walrasian output. Then prices would be high, and all firms would have profits, but since the firm producing the Walrasian output is producing more, its profits would be higher, and therefore imitated. In the V–R economy, firms almost always imitate the most successful firm in the past round, making every firm to obtain the average profit. Time to time a firm mutates and this mutation can be either profit improving or not, if it does not increase profits, it is not imitated, if it increases profits, then all firms copy the new strategy and average profits increase in the next round. In our set up, using the average filter, those strategies that did not perform are eliminated, and cannot be chosen again. Only those mutations that improve with respect to the mean survive the elimination process, and the outcome is the same as in the V–R economy.

The second result is that using other filters rather than the average filter, the Walrasian allocation appears as a boundary in the behaviour of the firms. This is the case of the minimum profit filter used with large marginal costs and depicted in Fig. 6.3. In this particular case the Walrasian outcome appears as an upper bound of the stable strategy set for all firms. We have also studied the behaviour of this artificial economy in situations where the Walrasian allocation does not exist, as it is the case when fixed costs are present. In this situation we have also seen that the behaviour that in a regular economy would have selected the Walrasian allocation, generates the profits predicted by the Walrasian allocation and sometimes more than that. When we use the zero profit filter, that is, firms delete from the strategy set those strategies that did not yield a positive profit, we observe again, that any strategy strictly higher than zero and below the Walrasian output can be randomly selected. Again, the Walrasian output is found as an upper bound in the behaviour of firms.

Results obtained from experiments suggest that real agents could be behaving according to a rule that takes into account relative profits, because simple agents that filter out those strategies that do not yield at least average profits replicate the outcome. And this simple behaviour is consistent with Vega-Redondo (1997) economy.

The paper shows how extremely simple agents can obtain, in certain circumstances, the Walrasian equilibrium, in finite time with no memory. This simplicity comes with some cost. Learning is, for example, difficult to introduce in this context. Since these simple agents delete strategies that did not perform according to some criteria at each iteration, those deleted strategies are no longer available, even when they could prove to be better in case that some exogenous environmental parameter changes like, for example, the slope of the demand curve or the number of firms in the market.

Acknowledgements G. Fernández-de-Córdoba acknowledges financial support from research grant SEJ2007-66592-CO3-03. This paper was motivated by C. Alos-Ferrer to whom I am indebted.

References

Antoni Boshc-Domenech and Nicolaas J. Vriend. Imitation of successful behavior in cournot markets. *The Economic Journal*, 113:495–524, 2004.

Edmund Chatoe. Just how (un) realistic are evolutionary algorithms as representations of social processes? *Journal of Artificial Societies and Social Simulation*, 1(3), 1998. URL `http://www.soc.surrey.ac.uk/JASSS/1/3/2.html`.

David Chavalarias. Metamimetic games: Modelling metadynamics in social cognition. *Journal of Artificial Societies and Social Simulation*, 9(2), 2006. URL `http://www.soc.surrey.ac.uk/JASSS/9/2/5.html`.

Gonzalo Fernández-de Córdoba. Walrasian and marshallian equilibrium compared by means of evolutionary rules with no imitation. Mimeo.

Vernon L. Smith. Theory, experiment and economics. *Journal of Economic Perspectives*, 3(1): 151–169, 1989.

Fernando Vega-Redondo. The evolution of walrasian behaviour. *Econometrica*, 65(2):375–384, 1997.

Chapter 7
Multidimensional Evolving Opinion for Sustainable Consumption Decision

Sabine Garabedian

Abstract This paper proposes a consumption model where the key of the purchase decision is the agent opinion. This opinion is multidimensional because it is shaped by direct opinion which depends on the agent's perception and by indirect opinion which depends on social interaction. This framework is applied to the purchase decision of ethical goods to study the evolution conditions to improve the sustainable consumption. So, for testing the diffusion of ethical opinion on the population, we have built an agent-based model named Level of Ethical Opinion (LEO). This model simulates the evolution of opinion when the agents discuss (strong interaction) and when they change their good perception following information modification (weak interaction).

7.1 Introduction

The development of sustainable consumption which is empirically observable is hard to explain. Indeed, the traditional literature does not suggest how the preferences used in the choice procedure are actually developed. Standard economic theory assumes perfectly rational agents acting upon exogenously given preferences. However, in experimental studies the process of creating preferences is widely recognized to be influenced by the context in which they are applied (Valente, 2003).

Since the development of the consumer society, some authors have highlighted the social and historical significance of purchase decisions. The conspicuous consumption theory developed by Veblen (1899) was the pioneer work in this field. This theory explains that goods and services are acquired mainly for the purpose of displaying income or wealth. In the mind of the conspicuous consumer, such display serves as a means of attaining or maintaining social status. Other authors as

---

S. Garabedian
GREDEG, University of Nice Sophia-Antipolis-CNRS, 250 av. Albert Einstein, 06650 Valbonne, France, e-mail: garabedi@gredeg.cnrs.fr

Leibenstein (1950) or Corneo and Jeanne (1997), take up this concept to describe the imitation or bandwagoning phenomenon. In this analysis, the preferences are still exogenous and the consumers are perfectly rational.

More recently, some evolutionary works develop the idea that consumption choice depend on pattern of past consumption (Bianchi, 1998; Langlois and Cosgel, 1998; Witt, 2001) through the notion of "*path-dependency*". The habits create some behaviouristic routine which lock the consumer in consumption style. In this assumption of endogenous preferences formation, the process of creating preferences is widely recognized to be influenced by the context in which they are applied: "[...] *there is a growing body of evidence that people's preferences depend on the context of choice, defined by the set of options under considerations*" (Shafir et al., 1993, p. 21). Here, the rationality is no longer perfect but bounded. More precisely, the rationality is "*situated*" in the context in which consumers make their decisions (Orlean, 1995; Lawson, 1997). This situated rationality inspired by Simon's (1955) work and detailed by Walliser (2000), concerns rationality that is constructed thought interaction and that involves rationally adaptive agents.

So, we consider a evolutionary approach of decision theory where the opinion which is the heart of decision-making, evolves in time. In situated rational representation of consumers, this opinion has both social and historical dimensions. This approach draws attention to the diffusion process to explain the evolution of behaviors. The fundamental hypothesis of this notion of diffusion is that individual interactions are the main driving force behind the evolution of individual behaviors. This interaction can be direct, individual to individual, or indirect, i.e. relay by a media or an institution as newspaper, advertisement, political party, etc. (Steyer and Zimmerman, 2004).

In this way, we propose a model of diffusion of ethical opinion as a multidimensional process to explain the dynamic evolution of sustainable consumption behavior. Consumers purchase goods differentiated by an ethical characteristic which determines heterogeneous opinions according to social and historical motivations of consumers. These opinions evolve through both weak interactions (in reaction to some information modifications) and strong interactions (discussions). The model uses the results obtained in the cognitive sciences, implementing an algorithm representing situated rationality that is compatible with the substantial evidence on actual people's behaviour from experimental studies.

7.2 Multidimensional Opinion

The interaction between agents has an impact on the patterns of opinion diffusion. On the one hand, the strong interaction leads an indirect opinion by the opinion confrontation through discussions. On the other hand, the weak interaction leads a direct opinion through the perception variations due to the modifications of information level.

On the case of decision of sustainable consumption, we consider an opinion about an ethical characteristic which differentiate the goods. So, the opinion $\beta_i(x) \in [-1,1]$ of consumer i about good with ethical characteristic x evolve through the formation of both direct and indirect opinion.

We denote by t the time which is discrete. So, the dynamic process of evolution can be described as follows:

- First, at each step, consumer opinion evolves through discussion with the other consumers. The evolution of opinion depends on both consumer opinion at the step t ($\beta_i(x,t-1)$) and other consumers opinions at this same step ($\beta_j(x,t-1)$). This is the indirect opinion.
- Second, at each $\rho > 1$ step, the consumer opinion evolves with some exogenous modification of information (I). This evolution depends on both perception variation which are represented by the opinion elasticity at information ($\varepsilon_{\beta_i(x)/I}$), and the level of consumer opinion at the modification step ($\beta_i(x,t-1)$). This is the direct opinion.

All in all, the algorithm used to determine a the opinion of i on good x at the step T is given by:

$$\beta_i(x,T) = \sum_{t=1}^{T} h(\{\beta_j(x,t-1)\},\beta_i(x,t-1))_{j\neq i} + \sum_{t=1}^{T} g\big(I,\beta_i(x,t-1),\varepsilon_{\beta_i(x)/I}\big).\theta(t) \tag{7.1}$$

where

$$\theta(t) = \begin{cases} 1 & \text{if } t \text{ is a multiple of } \rho \\ 0 & \text{otherwise} \end{cases}$$

where $h(.)$ gives the indirect opinion leaded by relative agrement model and $g(.)$ gives the direct opinion leaded by the perception variations when the information level changes.

7.2.1 Direct Opinion: An Opinion About the Characteristic

Some authors have discussed how new wants stem from the general economic growth (Witt, 2001), or how purchasing behaviour is influenced by social considerations (Cowan et al., 1997; Aversi et al., 1999). The direct opinion explains consumer opinion about a good with ethical characteristic. As ethical characteristic is observable through a label (*AB*, *Fair Trade*, etc.), this opinion depends on information level on the sustainable development. Indeed, some campaign of information about condition productions can modify this opinion. We suppose that there is an information-elasticity of the opinion $\varepsilon_{\beta(x)/I} \in [0,1]$ which determines the variation of the opinion after a information modification.

So, the function $g(.)$ which defines the state of opinion $\beta_i(x,t)$ of consumer i on characteristic x at the step t after a information modification $I \in [0,1]$ in step $[t-1]$, can be written as:

$$g : \beta_i(x,t) = \beta_i(x,t-1)\big(1 + \beta_i(x,t-1).\,\varepsilon_{\beta(x)/I}.\,I\big) \tag{7.2}$$

where $\beta_i(x,t-1).\,\varepsilon_{\beta(x)/I}.\,I$ explain the variation proportion of the opinion after an information variation I.

The formation of direct opinion and its evolution are based on social and historical motivation. Social motivation deals with a need of demonstration through the Veblen effect (a consumer considers good only), a need for membership through the bandwagon effect when consumer opinion about one good is negatively correlated with the number of consumers of the same good, or the imitation effect when this correlation is positive. Historical motivation refers to routines which are developed by evolutionist trends. In this way, the agents have more or less aversion to change their opinion. This aversion is given by uncertainty which obstructs change.

To take into account this two dimensions we propose an opinion defined by a segment where the social motivation gives the position of the segment $\beta_i(x)$, and the historical dimension which appear through uncertainty $\alpha_i \in [0,1]$ gives the length of the segment (Eq. 7.3).

$$S_i = [\beta_i(x) - \alpha_i, \beta_i(x) + \alpha_i]. \tag{7.3}$$

So, information-elasticity of the opinion $\varepsilon_{\beta(x)/I}$ takes into account both elasticity of opinion for social motivation $\varepsilon_{\beta(x_s)/I}$ (a value for each type of behavior: Veblen effect, bandwagon effect and imitation effect) and uncertainty for historical motivation α_i (a value for each agent). We write:

$$\varepsilon_{\beta_i(x)/F_k} = \varepsilon_{\beta_i(x_s)/F_k}.\,\alpha_i. \tag{7.4}$$

De facto, the values of the information-elasticity of opinion have an influence on the pattern of opinion diffusion.

7.2.2 *Indirect Opinion: An Opinion Resulting from Social Interaction*

The social interaction leads to convergence of opinion as shown by numerous works on opinion dynamics (Arthur, 1989; Orlean, 1995; Latané and Nowak, 1997; Deffuant et al., 2002; Urbig and Malitz, 2005). Each consumer has an indirect opinion on the good which depends on his interaction with an interlocutor. This interaction leads to a relative agreement model similar to the one in Deffuant et al. (2002). In this model, the agents have continuous opinions $\beta \in [-1,1]$ and uncertainties $\alpha \in [0,1]$ about this opinion. At each step, there is an interaction between a random pair of agents. If we consider two agents i and j, the change in opinion β_i of agent i under the influence of agent j is proportional to the overlap between both segments (the agreement), divided by the uncertainty of the influencing segment (which explains why we call it "relative"). After agents i and j's opinion updating, a new pair is randomly chosen and the same process is iterated (Deffuant et al., 2002).

Formally, the agreement of agent j with i is defined by the overlap of S_j and S_i, minus the non-overlapping part.

The overlap h_{ij} is given by:

$$h_{ij} = \min(\beta_j + \alpha_j, \beta_i + \alpha_i) - \max(\beta_j - \alpha_j, \beta_i - \alpha_i).$$

The non-overlapping width is:

$$2\alpha_j - h_{ij}.$$

The agreement is the overlap minus the non-overlap:

$$h_{ij} - (2\alpha_j - h_{ij}) = 2(h_{ij} - \alpha_j).$$

The relative agreement is the agreement divided by the length of segment S_j.

$$\frac{2(h_{ij} - \alpha_j)}{2\alpha_j} = \frac{h_{ij}}{\alpha_j} - 1.$$

If $h - ij > \alpha_j$, then the modification of β_i and α_i by the interaction with j is:

$$\Delta\beta_i = \left(\frac{h_{ij}}{\alpha_j} - 1\right)(\beta_j - \beta_i). \tag{7.5}$$

A feature of this model is an asymmetric influence of agents, when they have different uncertainties. The confident agents (with low uncertainty) have greater influence. This corresponds to the common experience in which confident people tend to convince uncertain people (when the differences in opinion are below a certain threshold).

7.2.3 Consumers Classification

Moreover, we suppose that consumers are heterogeneous in their socio-historical motivations and in their propensity to discuss. This heterogeneity ensures classified consumers into three social groups: proactive, median, nonconformist. This segmentation has two effects on the way in which both direct and indirect opinion evolve. Depending on the consumer membership, the perception variations are different and his discussions are more or less frequent.

The consumers will not have the same behavior and the same behavior evolution due to their different social and historical motivations. Depending on the social and historical motivations, we can define three type of consumer: a *proactive group* which leads a Veblen effect and which is confident (uncertainty weak), a *median group* which gives rise to imitation effects and which is moderately confident, and

Table 7.1 Consumers classification in groups

α_i \ $\beta_i(x)$	$\in [-1, \overline{\beta}]$	$\in [\overline{\beta}, 1]$
$\in [0, \overline{\alpha}]$	Median	Proactive
$\in [\overline{\alpha}, 1]$	Median	Nonconformist

a *nonconformist group* which leads a bandwagon effect and which is not confident (uncertainty strong). To classify the consumers according to their values of opinion segment compared with the middle values, we denote by $\overline{\beta}$ the middle opinion and by $\overline{\alpha}$ the middle of uncertainties.

For all groups, the purchase decision of ethical good depends on ethical opinion level. So, for the proactive group, opinion is higher than the middle opinion because the consumer who wants to flaunt his ethical value will have a good opinion on the ethical characteristic.

For median and nonconformist, the opinion depends on the number of consumers who purchase ethical goods (i.e. who have a opinion about the ethical characteristic higher than the middle opinion). As the ethical goods are not widely distributed, the median group has an opinion lower than the middle opinion and the nonconformist have an opinion higher than the middle opinion.

The proactive and nonconformist groups have both a opinion higher than the middle opinion but they are different by their uncertainties. Indeed, the proactive group is convinced about his opinion when the nonconformist group is not to be sure about it. So, we can classify the consumer as shown in Table 7.1

7.3 Computer Simulation and Results

To implement the theoretical model, we have constructed an agent-based model denoted by LEO (Level Ethical Opinion), which simulates a diffusion of ethical opinion. The starting point of LEO is the works of Deffuant et al. (2002) and Amblard (2003) which developed a model to analyse the propagation of innovation (Deffuant, 2001) and the agri-environmental measures adoption by farmer (Deffuant et al., 2002). LEO include their assumption of interaction, i.e. the model of relative agreement to shape the indirect opinion, and we add two other mechanisms: first, an institution modifies positively the information in the course of discussions to shape the direct opinion, and second, the population is segmented in different group according to the positioning of the opinion segment. Then, the evolution process (direct and indirect) is modulated depending on the group membership.

We have tested the effects of two elements. Firstly, we have looked at the influences of both weak and strong sensibility to information for the proactive group. Secondly, we have considered, through the different discussion rate for median and nonconformist groups, the effects of communication between groups on the patterns of opinion diffusion.

We have simulated 10 replications (set of similar parameters on different populations) for 18 series (set of different parameters) to test the two propositions. Altogether, 720 simulations have been made.

7.3.1 Groups' Characteristics

The population segmentation has two effects on the way which both direct and indirect opinions evolve.

On the one hand, opinion-elasticity ($\varepsilon_{\beta_i(x)/I}$) which guides the evolution process after information modification, is related to group membership. The proactive group has a positive elasticity because its consumers lead a Veblen effect. We test the influence of two levels of elasticity: 0,5 and 1. The median and nonconformist groups have an elasticity linked to the others' opinion because its consumers lead an imitation and a bandwagon effect, respectively. Table 7.2 shows the value of information-elasticity of opinion.

On the other hand, the frequency of discussion which appears through the discussion rate $\chi \in [0,1]$, is related to group membership. This discussion rate explains the proportion of the consumers social network which is activated as interlocutors for the discussions. Indeed, among his social network, a consumer can talk about sustainable development with only some of his acquaintances. The selection of the interlocutors is shown in Fig. 7.1. So, we suppose that the proactive group has a discussion rate maximum due to his display motivation ($\chi = 1$), and the discussion rate of the two other groups will be tested with the computer simulation model. Three values are tested: $\chi = 0.2, 0.5, 0.8$ to experiment influence of discussions respectively less, medium, and high frequency, on the pattern of opinion diffusion.

Table 7.2 Values of information-elasticity of opinion for social motivation ($\varepsilon_{\beta(x_s)/I}$) for each group

Group \ Information	high	low
Proactive	1	0.5
Median	$\overline{\beta}$	$\overline{\beta}$
Nonconformiste	$-\overline{\beta}$	$-\overline{\beta}$

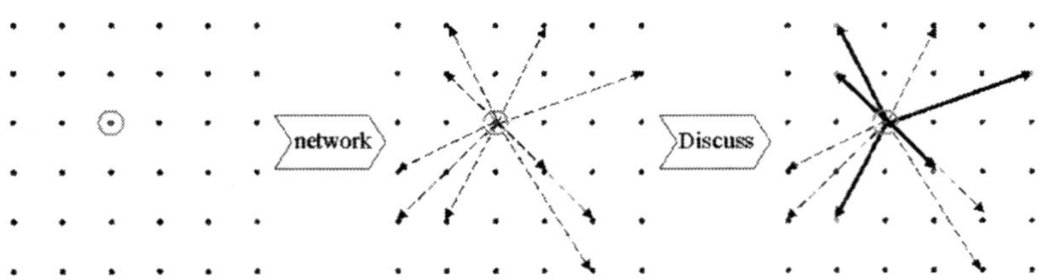

Fig. 7.1 Social network and discussion rate

7.3.2 Impact of Elasticity Values

When new information is sent, the middle opinion leaps. This new level of opinion is maintain by the discussions. Figure 7.2 exposes the pattern of middle opinion evolution with 10 positive modifications of information for each 10 replications when the elasticity is 1 (left) and 0,5 (right).

The left figure presents the case where the value of information-elasticity of opinion is 1. In this case, all replication simulations show a homogeneous growing opinion. The right figure presents the case where the value of information-elasticity of opinion is 0,5. In this case, the replication simulations show heterogeneous series where some trajectories are locked.

This results shows that adoption of new opinion is not systematic even if there is an information campaign. It corroborates some evolutionist works on technology diffusion. David (1985), Arthur (1989) or Liebowitz and Margolis (1994) highlighted the presence of technological lock-in the diffusion of competing technologies on the basis of theoretical concepts developed by Nelson and Winter (1982) or Dosi (1988). LEO simulations show that lock-in effect is also present for the consumption style when the information-elasticity of opinion is not enough. On the other words, if the consumers are not swayed enough by the information, they don't revise their scale of value which allows a positive evaluation of ethical characteristic.

Moreover, we can observe that locked-in trajectories reveal similarity in the repartition structure of population in groups. To examine this repartition, we analyse the influence of the discussion rate.

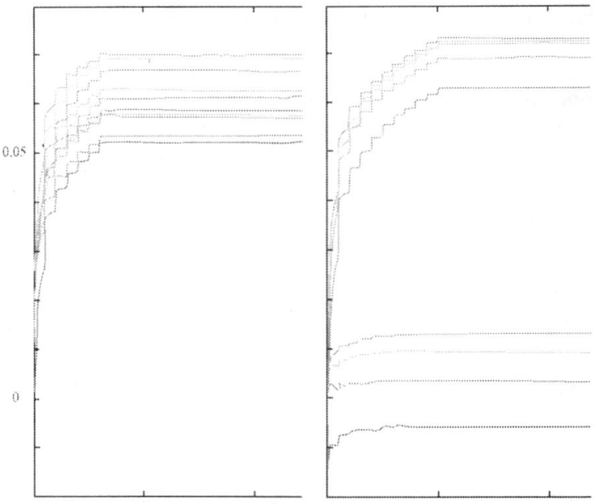

Fig. 7.2 Variation of middle opinion for 10 replicas for $\varepsilon_{\beta(x_s)/I} = 1$ (*left*) and $\varepsilon_{\beta(x_s)/I} = 0,5$ (*right*)

7.3.3 Impact of Discussion Rate

The discussion rate have an impact on the number of other consumers with which the consumer talks. Thus, a weak discussion rate for a particular group means a weak connectivity for its group members. So, we can observe that locked-in trajectories occur when the discussion rate about the median group is weak ($\chi = 0,2$). In this case, the information modifications are not transmitted by discussions. So, the pattern of diffusion of opinion can be locked.

This weak discussion rate creates a specific evolution pattern of population repartition on the different groups. Indeed, in Fig. 7.3 which shows the evolution in time of groups size, we note that in the case of homogenous increasing middle opinion (left), the proactive group grows in the first simulation periods: some median consumers became proactive consumers. However, in the case where there is lock-in trajectories of opinion (figure right), there is no consumer transfer between the median and the proactive group. If this conversion does not take place in the first steps of simulations, the population cannot get under way the increasing process of opinion. In this case, the positive information modifications have no influences on the middle opinion.

This result agrees with several works on the learning in networks in non-strategic interaction (Allen, 1982; Ellison and Fudenberg, 1993, 1995; Bala and Goyal, 1998). This type of information flow (transmitted by discussion) may lead to efficient learning on the social level, but this efficiency is linked to a large extent to network connectivity. Here, a weak discussion rate, and so a weak connectivity, does not makes it possible to convince enough median consumers to get the opinion improvement process underway.

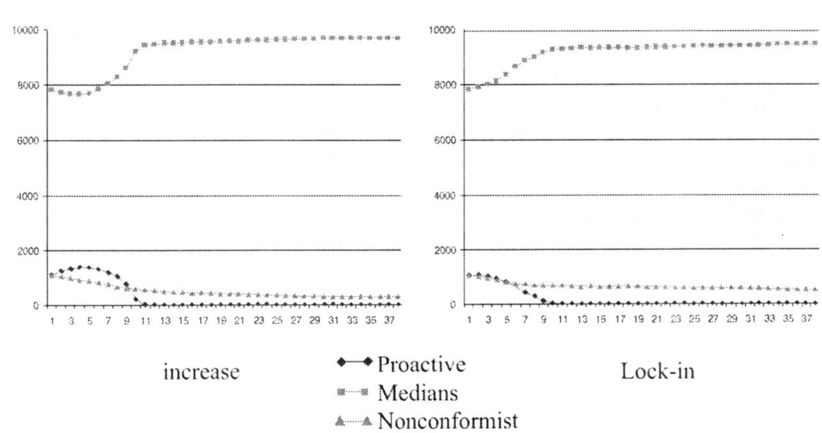

Fig. 7.3 Evolution of group size

7.4 Conclusion

The results obtained with such a model vary depending on value of information elasticities of opinion. When the consumers are not swayed enough by information, some lock-in situations can occur. These situations are characterized by a specific evolution of group size – no demographic growth of proactive group – due to a too weak discussion rate. This result shows that communication between the different groups is the critical point to promote a specific opinion.

In addition to these specific conclusions, we hope that our analysis demonstrates that situated rational models are an interesting and tractable way to understand some aspects of opinion diffusion.

References

B. Allen. Some stochastic processes of interdependent demand and technological diffusion of an innovation exhibiting externalities among adopters. *International Economic Review*, 23:595–607, 1982.

F. Amblard. *Comprendre le fonctionnement de simulations sociales individus-centré*. PhD thesis, Université Blaise Pascal, Clermont II, 2003.

J. B. Arthur. Competing technologies, increasing returns, and lock-in by historical events. *The Economic Journal*, 99:116–131, 1989.

R. Aversi, G. Dosi, G. Fagiolo, M. Meacci, and Olivetti C. Demand dynamics with socially evolving preferences. *Industrial and Corporate Change*, 8:353–408, 1999.

V. Bala and S. Goyal. Learning from neighbours. *Review of Economic Studies*, 65:595–621, 1998.

M. Bianchi. *The active consumer*. Routledge, London, 1998.

G. Corneo and O. Jeanne. Conspicuous consumption, snobbism and conformism. *Journal of Public Economics*, 66:55–71, 1997.

R. Cowan, W. Cowan, and P. Swann. A model of demand with interaction among consumers. *International Journal of Industrial Organization*, 15:711–732, 1997.

P.A. David. Clio and the economics of qwerty. *American Economic Review*, 75(2):332–337, 1985.

G. Deffuant. Improving agri-environmental policies: A simulation approach to the cognitive properties of farmers and institutions. Technical report, Final report of project FAIR 3 CT 2092, 2001.

G. Deffuant, F. Amblard, G. Weisbuch, and T. Faure. How can extremism prevail? a study based on the relative agreement interaction model. *Journal of Artificial Societies and Social Simulation*, 5(4), 2002.

G. Dosi. Sources, procedures, and microeconomic effect of innovation. *Journal of Economic Literature*, 26(3):1120–71, 1988.

G. Ellison and D. Fudenberg. Word-of-mouth communication and social learning. *Quarterly Journal of Economics*, 109:93–125, 1995.

G. Ellison and D. Fudenberg. Rules of thumb for social learning. *Journal of Political Economy*, 101:612–644, 1993.

R. N. Langlois and M. M. Cosgel. The organization of consumption. In *The active consumer*. Routledge, London, 1998.

B. Latané and A. Nowak. Self-organizing social systems: Necessary and sufficient conditions for the emergence of clustering, consolidation, and continuing diversity. In G. A. Barnett and F. J.

Boster, editors, *Progress in Communication Sciences: Persuasion*, volume 13, pages 43–74. Norwood, NJ, 1997.

T. Lawson. Situated rationality. *Journal of Economic Methodology*, 4(1):101–125, 1997.

H. Leibenstein. Bandwagon, snob and veblen effects in the theory of consumers' demand. *The Quarterly Journal of Economics*, 64(2):181–207, 1950.

S. J. Liebowitz and S. E. Margolis. Network externality: An uncommon tradegy. *Journal of Economic Perspectives*, 8(2):133–150, 1994.

R. R. Nelson and S. G. Winter. *An Evolutionary Theory of Economic Change*. Harvard University Press, 1982.

A. Orlean. Bayesian interactions and collective dynamics of opinions: Herd behavior and mimetic contagion. *Journal of Economic Behavior and Organization*, 28:257–274, 1995.

E. B. Shafir, I. Simonson, and A. Tversky. Reason-based choice. *Cognition*, 49:1–36, 1993.

H. A. Simon. A behavioral model of rational choice. *Quarterly Journal of Economics*, 69:99–118, 1955.

A. Steyer and J. B. Zimmerman. Influence sociale et diffusion de l'innovation. *Mathématiques et sciences humaines*, 168:43–57, 2004.

D. Urbig and R. Malitz. Dynamics of structured attitudes and opinions. In K. G. Troitzsch, editor, *Representing Social Reality, Pre-Proceedings of the Third Conference of the European Social Simulation Association (ESSA)*, pages 206–212, 2005.

M. Valente. Simulation methodology: an example in modelling demand. Universit'a dellAquila, 2003.

T. Veblen. *Théorie de la classe de loisir*. Gallimard, 1899.

B. Walliser. *L'Economie Cognitive*. Odile Jacob, 2000.

U. Witt. Learning to consume - a theory of wants and the growth of demand. *Journal of Evolutionary Economics*, 11(1):23–36, 2001.

Part III
Information Economics

Chapter 8
Local Interaction, Incomplete Information and Properties of Asset Prices

Richard Hule and Jochen Lawrenz

Abstract In this article, we introduce local interaction in a pure exchange economy where the endowment process follows a simple hidden Markov chain and risk-averse agents have incomplete information about the regime. We show that the interplay between internal, external local and external global effects (a) can account for different temporal behavior of the price-dividend ratio, (b) can reproduce some stylized facts of price changes and (c) suggest that market efficiency in the sense of return predictability may be inversely related to the information precision.

8.1 Introduction

In the idealized world of neoclassical economics, agents are most frequently assumed to possess full information about the state of the economy and that their decisions can be analyzed by those of a single representative agent. It is fair to say that real-world agents are less than perfectly informed and that they reveal some kind of heterogeneity.[1] Studying the outcome of financial markets if the representative agent assumption is dropped, is a central theme of work known under the notion of agent-based, interacting agents, or heterogenous agent models.[2]

[1] Without trying to discuss this in detail, it should be noted, that an advocate of neoclassical economics arguably does not claim that all agents *are* perfectly informed and identical, but that it is reasonable to analyze the aggregate outcome *as if* these assumptions were justified.

[2] For reviews on agent-based models, see e.g. Hommes (2006) or LeBaron (2006).

R. Hule
Department of Economics, Innsbruck University, Universitätsstrasse 15, 6020 Innsbruck, Austria, e-mail: richard.hule@uibk.ac.at

J. Lawrenz
Department of Banking & Finance, Innsbruck University, Universitätsstrasse 15, 6020 Innsbruck, Austria, e-mail: jochen.lawrenz@uibk.ac.at

A drawback of the representative agent methodology is that one inevitably has to abstract from (almost) any 'social' interactions between agents.[3] It is true that in our modern world a large part of information and interaction is done quite anonymously by using market institutions and the prices generated by them. Nevertheless, especially in an uncertain environment, markets are based on the behavior of individual agents, which definitely takes place in a social context and not in splendid isolation. Arguably, social interactions may be more important if the quality of information is rather bad. Intense social interactions may then lead to herd behavior and other positive feedback mechanism, which, in turn, will have an influence on the dynamics of asset prices. The goal of this paper is to contribute to the understanding of these dynamics by analyzing a simple agent-based model with enough structure to investigate the impact of the information quality and the intensity of local interaction on asset price characteristics, such as time series properties of the price-dividend ratio, return predictability or the risk premium.

Obviously, we are not the first to address some of these issues. There exists an extensive literature on different aspects, that we try to link in our contribution. We will shortly review part of this literature in order to point out connections to our model and also to motivate some of the modelling choices. The main contribution of the paper can be seen as modelling the feedback between social interaction, quality of information and properties of asset prices. In a broader sense, our contribution is another step in the attempt to bridge the gap between the 'classical' finance literature and the modelling approach put forward by the more recent agent-based literature.

A large body of work in 'classical finance' relies on the representative agent paradigm and perfectly rational utility maximization.

Early work has focused on the impact of incomplete information on asset pricing and portfolio allocation. Seminal contributions are due to Detemple (1986), Dothan and Feldman (1986), or Gennotte (1986). More recently, Veronesi (1999), Veronesi (2000), or Brennan and Xia (2001) address the issues of overreaction, excess volatility and the equity premium puzzle within related setups.

An important feature of these models is, that in contrast to models like those in e.g. Cont and Bouchaud (2000) and other related work in econophysics, they consider sophisticated market mechanism and equilibrium concepts, where the demand of investors is determined by maximizing expected utility. Therefore, they are able to draw conclusions on equilibrium asset returns and related risk premia.

With respect to our basic research question, our model is most closely related to Veronesi (2000) and Brennan and Xia (2001). Both contributions study the impact of incomplete information on the properties of stock returns, in particular on the stock price volatility and the equity premium. Brennan and Xia (2001) find that their model is able to explain excess volatility as well as a high equity premium. However, they have to assume a rather high risk aversion parameter to obtain these results. Veronesi (2000) on the other hand finds that the incomplete information makes the equity premium puzzle even more puzzling, since he shows that a lower precision of signals (i.e. worse information quality) tends to decrease the risk premium.

[3] See e.g. Kirman (1992) for a critique on the representative agent assumption.

rather than increasing it. Veronesi (2000) as well as Brennan and Xia (2001) consider a pure Lucas exchange economy with two securities: a risk-free and a risky asset. The risky asset pays a stochastic dividend, following a standard diffusion process. Agents can observe the dividend flow, but cannot observe the drift rate. While Brennan and Xia (2001) assume that the drift rate follows an Ornstein–Uhlenbeck process, in Veronesi (2000), the drift rate can assume only finitely many values. More importantly, in Brennan and Xia (2001), agents have to infer the drift rate only from observations on the dividend flow, while in Veronesi (2000), they receive an additional noisy signal. In their rational expectations setup, the focus is on processing the information by Bayesian updating, and thus their optimal inference problem is solved by applying Kalman–Bucy filter results.[4] Our model will basically follow a similar approach by considering a Lucas economy with a risky asset, whose drift rate may switch with some probability between high and low. As in Veronesi (2000), all agents will receive a noisy signal about the drift rate in addition to the information included in the dividend of each period. If we allow for local interaction in a finite-agent economy, the complexity of the optimization problem (i.e. basically the Euler equations of individual agents) cannot be resolved analytically, and equilibrium results can only be obtained by numerical procedures.

With respect to the aspect of local interaction, our model is related to contributions by Orlean (1995), Stauffer and Sornette (1999) or Cont and Bouchaud (2000). In Orlean (1995), agents update their beliefs not only on the basis of their private signal, but also by observing the collective "group opinion", which aggregates individual decisions. He thereby generalizes the herding results[5] to a non-sequential setting. The model by Cont and Bouchaud (2000) for example, has gained much recognition as a simple model that is able to produce stylized market facts that are consistent with empirical observations. They allow for a heterogenous market structure, where different groups can make independent decisions. The formation of groups is assumed to be random, and the formal tool which they apply is random graph theory.

We take up the basic intuition from these models to formalize the idea of social interaction. Obviously, there are a lot of modelling choices to be made at this stage,[6] and most of them will be made to keep the model simple and tractable. First, we choose a fixed network (see Wilhite (2006)), basically a cellular automaton,[7] rather than endogenous formation of the links between the neighbors.[8] Second, we do not assume that the state of each agent mechanically determines his demand, but rather this feeds back into his demand decision. Third, in contrast to several models (e.g. Lux and Marchesi (1999) or Farmer and Joshi (2002)), we do not assume

---

[4] For a concise review on the application of filtering theory in asset pricing, see e.g. Back (2004).

[5] See e.g. Bikhchandani et al. (1992) for the classical sequential herding model.

[6] See LeBaron (2006) for a discussion of modelling choices.

[7] See Wolfram (2002) on cellular automata.

[8] This amounts to saying something about the relative speed of change of trading and social interactions. Namely, we assume that the social environment is stable and changes more slowly than the dividend process of the risky asset (see e.g. Vriend (2006)).

agents that act as chartists or trend followers, since in this way an automatic positive feedback mechanism is already model-built-in right from the outset.

The next section sets up the model. In Sect. 8.3, simulation results are presented and discussed. A summary concludes the article.

8.2 The Economy

In line with a substantial part of the finance literature, we consider a pure-exchange economy in the spirit of Lucas (1978). There exists an exogenous stochastic non-storable endowment process which is traded in the form of one unit of stock, being infinitely divisible. We assume time to be discrete with an infinite horizon. At each point in time t, the stock pays a dividend whose dynamics follow the discretized version of a standard diffusion process, i.e.

$$D_t = D_{t-\Delta t} \exp\left\{ \left(\mu - \frac{\sigma_D^2}{2} \right) \Delta t + \sigma_D \sqrt{\Delta t} \varepsilon_t \right\} \tag{8.1}$$

where the ε_t are standard normal iid, and σ_D is constant.[9] Similar to Veronesi (2000) and Brennan and Xia (2001), we assume that the growth rate μ can take on two possible values $\mu \in \{\mu_H, \mu_L\}$ with $\mu_H \geq \mu_L$, whose actual realization is not observable by any agent. At any point in time, there is a fixed probability p_μ that the growth rate changes. Thus p_μ can be thought of as the probability that the economy shifts into the other regime.[10] While all parameters are assumed to be common knowledge and agents cannot observe the actual growth rate, i.e. the current regime, they receive a signal about μ, denoted by e, which is for all t a noisy version of μ_t, i.e.

$$e_t = \mu_t + \sigma_e \eta_t. \tag{8.2}$$

Again, η_t is standard normal iid, and σ_e is a constant. The information contained in observing the process (D_t, e_t) can be used to draw inference about μ. Since (D_t, e_t) is available to all agents, we assume that every agent arrives at the same estimate for the economy to be in regime H, which we denote as $\hat{\mu} = Prob\{\mu_t = \mu_H | (D_s, e_s)_{s<t}\}$.[11] Since our model setup can be characterized as a regime-switching model (sometimes also called Hidden Markov models) with Gaussian innovations,

---

[9] In general, the agents opportunity set is frequently assumed to additionally contain a riskless asset which is in zero net supply. However, note that in this endowment economy, equilibrium requires that aggregate consumption equals aggregate dividends, so that all agents only hold the risky asset, and the rate of return on the riskless asset is only to be interpreted as the economy's shadow riskfree interest rate.

[10] For convenience, this is assumed to be symmetric. i.e. $Prob\{\mu_t = \mu_H | \mu_{t-1} = \mu_L\} = Prob\{\mu_t = \mu_L | \mu_{t-1} = \mu_H\}$.

[11] Alternatively, suppose that estimates are made by a central authority in the economy, say in the form of the economic outlook by the central bank, which then is available to all agents.

the inference problem can be solved by applying a variant of the Kalman filter.[12] Denote by $n_{i,t}$, the joint (bivariate) density of the innovations in regime i at time t,[13] then for any t, the updated probability $\hat{\mu}$ is

$$\hat{\mu}_t = \frac{n_H \left(p_\mu + \hat{\mu}_{t-1} - 2p_\mu \hat{\mu}_{t-1}\right)}{n_L + (n_H - n_L)\left(p_\mu + \hat{\mu}_{t-1} - 2p_\mu \hat{\mu}_{t-1}\right)}.$$

Note that $\hat{\mu}$ is the estimated probability for regime H, and not the estimate for μ itself. To set up the model, we basically have to consider three broad issues: The first is the social interaction mechanism, the other two are the (boundedly rational or adaptive) formation of price expectations and the demand, which then determines the market clearing price for the risky asset.

Local interaction The fundamental aspect where we depart from the classic Lucas economy setup is that we drop the representative agents assumption and consider an economy with a finite number N of agents with local interaction. In order to give meaning to the notion of 'locality', we need to impose some topology on our economy. From the multitude of possible fixed networks, for the sake of simplicity, we choose the ring network.[14] Thus, we can think of the N agents to be situated along a line, where agent N is connected to agent 1. More formally, let us define the neighborhood of agent i as $\mathcal{N}_i = \{j;\, (i-1) \bmod N \leq j \leq (i+1) \bmod N\}$. Note that $\mathcal{N}_i$ is the set of the nearest adjacent agents on the ring, including agent i.[15] At every point in time t, we let every agent assume one out of two different *states*, denoted as $S_i(t)$.[16] Each agent can observe the state of the agents in his neighborhood. Therefore, he can determine the (collective) state of his neighborhood, which we denote as $S_{\mathcal{N}_i}(t)$.[17] The state of his neighborhood at time t will influence the state of the agent at the subsequent period $t+1$. If we only allowed for local interaction in this form, we would be in a pure contagion model as e.g. analyzed in Morris (2000).

---

[12] See e.g. Hamilton (2008).

[13] The joint distribution of innovations in the log dividend and signal process are bivariate normal, i.e.

$$\begin{pmatrix} \log D_t \\ e_t \end{pmatrix} \sim \mathcal{N}\left(\vec{\mu}_{i,t}, \Sigma\right), \qquad \text{where: } \vec{\mu}_{i,t} = \begin{pmatrix} \mu_{i,t} \\ \log(D_{t-1}) + (\mu_{i,t} + \sigma_D^2/2) \end{pmatrix}, \quad \Sigma = \begin{pmatrix} \sigma_e^2 & 0 \\ 0 & \sigma_D^2 \end{pmatrix}$$

[14] See Wilhite (2006) for a review of network topologies. In general, results will depend on the choice of the specific network. Morris (2000) however has shown that there exist conditions for which contagion is possible in arbitrary network topologies. Since we are in particular interested in contagion results, the specific choice of networks is not of primary importance for our purposes.

[15] On the one hand, it is uncommon to include the agent in his own neighborhood, since usually one assumes the condition of irreflexivity (i.e. no agent is a neighbor of his own), but on the other hand it has the advantage to rule out cyclical behavior which occurs if odd agents are in state 1 while even agents are in state 0.

[16] An extension to finitely many different states would be straightforward.

[17] Different choices are possible, but for simplicity, define the state of the neighborhood as the equally weighted average of the states of the agents in the neighborhood, i.e. $S_{\mathcal{N}_i} = |\mathcal{N}_i|^{-1} \sum_{k \in \mathcal{N}_i} S_k$.

However, we are interested in the interplay between local interaction and incomplete information about the regime of the economy. Therefore the state of a single agent will also be influenced by the estimate of the regime $\hat{\mu}$, derived from observing the dividend and signal process.[18] Thus, at an abstract level, we can summarize the dynamic evolution of the local interaction as the following updating rule,

$$S_i(t) = f\left(S_{\mathcal{N}_i}(t-1), \hat{\mu}(t), \phi\right). \tag{8.3}$$

The state of agent i at time t will depend on the state of his neighborhood (including his own state) at the previous time, and the current estimate about the regime. ϕ summarizes the weights given to the different influencing factors and will be crucial for our further analysis. Amongst others, ϕ controls for the influence of the neighborhood on a single agent, i.e. the intensity of local interaction.[19]

To make (8.3) concrete, we assume the following functional form,

$$S_i(t) = \begin{cases} 0 \text{ with probability } (1-\pi_t) \\ 1 \text{ with probability } \pi_t \end{cases},$$

where the probability π_t is given as

$$\pi_t = g\left(\phi_{\mathcal{N}} \mathcal{N}_i(t-1) + \phi_\mu \hat{\mu}(t)\right), \tag{8.4}$$

$$g(y) = \frac{1}{2}\operatorname{sgn}(2y-1)\,|2y-1|^{\phi_g} + \frac{1}{2}, \quad \text{for } 0 \le y \le 1, \tag{8.5}$$

with $\phi_{\mathcal{N}} + \phi_\mu = 1$ and $\phi_g > 0$. First, note that $\mathcal{N}_i$ as well as $\hat{\mu}$ are bounded between 0 and 1. The function g, which has some similarity to sigmoid functions, ensures that $y \in [0,1]$ is mapped to the unit interval, whereby the actual mapping depends on the parameter ϕ_g. For $\phi_g \to 0$, $\pi \to 1_{\{\phi_{\mathcal{N}} \mathcal{N}_i(t-1)+\phi_\mu \hat{\mu}(t)>1/2\}}$, while for $\phi_g \to \infty$, $\pi \to 1/2$. Thus the parameter ϕ_g controls for the individual randomness in the determination of the state of agent i. For high values of ϕ_g, the state is dominated by idiosyncratic, internal randomness. Low values of ϕ_g indicate that the state is predominantly determined by external influences. Note that there are two different external factors, whereby the state of the neighborhood is a local effect, while the regime-estimate is a global effect. Thus, we can succinctly summarize the controlling parameters in $\phi = \left(\phi_g, \phi_{\mathcal{N}}\right)$. ϕ will play a crucial role in the subsequent section, where we will analyze the impact of internal, external local and external global effects.

Price expectation, Demand, and Market clearing The state of an agent is assumed to determine if he expects the economy to be in the high or low growth regime, which, on the aggregate level, will have an impact on the market clearing

[18] Note that we do not consider the influence of market prices. There are basically two reasons: On the one hand, it is far from clear what agents can infer from market prices in such an environment, and on the other hand, in the pure exchange-economy prices have no speculative role.

[19] The interpretation is f.ex. equivalent to the parameter c in Cont and Bouchaud (2000), which they define as "the willingness of agents to align their actions".

price. For our purposes, it will be sufficient to assume that an agent i in state $S_i = 1$ expects the economy to be in the high-growth regime, while in state $S_i = 0$ he expects low growth rates. Loosely speaking, the two possible states might be interpreted as agents being pessimistic or optimistic, bearish or bullish, in good or bad mood or some similar dichotomy.[20]

Conditional on his state, agents in our endowment economy try to maximize their life-time utility from consumption. We assume that all agents have preferences that can be ordered by a time-additive utility function over consumption with constant relative risk aversion, i.e.

$$u(c_t) = \frac{c_t^{1-\gamma}}{1-\gamma}. \tag{8.6}$$

Thus, every agent tries to maximize $U(c) = E\left(\sum_{j=0}^{T} \beta^j u(c_{t+j})\right)$, where β represents the subjective time discount factor. In the sense of the Bellman equation, $U(c)$ can be decomposed in current utility of consumption (as of time t) and some expected future utility, where the latter is frequently denoted as the value function (V), i.e.

$$U(c) = u(c_t) + \beta V_{t+1}, \qquad \text{where: } V_{t+1} = E_t\left(\sum_{j=1}^{T} \beta^{j-1} u(c_{t+j})\right). \tag{8.7}$$

In general, V_{t+1} is supposed to be the optimum taken over all admissible consumption policies. In the present context, due to the complexity of the interaction, there is (at least to our knowledge) no way to provide an analytic solution to that optimization problem, and we are forced to assume some heuristic. We consider the case, where agents adopt a cautious or safety-first strategy in the following sense. For each t, agents consider their value function to consist of the future expected utility which they obtain if they keep their portfolio unchanged. Let $x_{i,t}$ denote the fraction of the risky asset that agent i holds at time t, then we take V_{t+1} to be $V_{t+1} = E_t\left(\sum_{j=1}^{T} \beta^{j-1} u(x_t \cdot D_{t+j})\right)$. To call this a cautious or safety-first strategy is justified, because this is the (expected future) utility, agents can lock in today. Therefore, it represents a minimum or lower bound on the value function.[21] Clearly, they could do better by taking into account their possibility to revise their portfolio in the future.[22] However, as noted above, it is far from clear how this can be optimally done.

---

[20] Considering the impact of the mood of investors on asset returns is by no means an exotic issue, but has already been discussed in the behavioral finance literature. Saunders (1993) and Hirshleifer and Shumway (2003) for example find evidence that the weather has an significant influence on stock returns, the channel being the mood of investors.

[21] This is also the reason why we ourselves were cautious in writing above, that agents *try* to maximize their life-time utility.

[22] Note, that since we assume agents to lock in their future consumption possibilities, their decision does not depend on future price changes, which is consistent with our assumption that the state on an agent will not be influenced by market prices.

Taking into account the dividend dynamics (8.1) and the utility function (8.6), it is easy to show that

$$V_{t+1} = \mathcal{V}_i \cdot u(x_t D_t), \qquad \text{where: } \mathcal{V}_i = \frac{\exp\{1/2(1-\gamma)(2\mu_i + \gamma\sigma_D^2)\}}{\exp\{1/2(1-\gamma)(2\mu_i + \gamma\sigma_D^2)\}\beta - 1}.$$

Note that the constant $\mathcal{V}_i$ depends on the growth rate μ_i. Finally, the budget constraint requires that for any t, consumption must be equal to the difference between the value of the portfolio before and after readjusting the asset holdings, i.e. let x_t^- be the fraction of the risky asset, i.e. the shares, right before readjustment, then $c_t = (P_t + D_t)x_t^- - P_t x_t$, where obviously, P_t is the market price of the risky asset. Plugging in the budget constraint and the value function in (8.7), the first-order condition for $\max_x U(c)$ is $u'((P_t + D_t)x_t^- - P_t x_t) + \beta \mathcal{V}_i u'(x_t D_t) = 0$, from which we find

$$x_{i,t} = \frac{\tilde{\mathcal{V}}_i x_t^- (1 + \omega_t)}{\tilde{\mathcal{V}}_i \omega_t + \omega_t^{1/\gamma}}, \tag{8.8}$$

where $\omega_t = P_t / D_t$ is the price-dividend ratio, and $\tilde{\mathcal{V}}_i = (\beta \mathcal{V}_i)^{1/\gamma}$. Note that we are able to explicitly derive a demand function for agents which depends on their state i. Furthermore, observe that the demand-price function is non-linear.

To close the model, let $X_{i,t}$ be the aggregate demand of agents in state i, then the market clearing condition requires $1 = \sum_i X_{i,t}$ for every t, from which the market clearing price can be determined numerically.

8.3 Simulation Results

As a consistency check, first note that if we reduce our setup to the single representative agent economy, then $x_{i,t} = x_t^- = 1$ and the price-dividend ratio simplifies to $\omega = \beta \mathcal{V}_i$, which is a constant and consistent with the one obtained by Mehra and Prescott (1985, 2003).

Now, if we consider our agent-based economy, results will crucially depend on ϕ, which governs the strength of internal vs. external and local vs. global influencing factors. In the first part of this section, we discuss results concerning the time-series properties of the price-dividend ratio for different constellations of ϕ.

Price-dividend ratio Although our focus is on qualitative results, we use the estimates reported in Mehra and Prescott (2003) for the required parameter to be grossly consistent with empirical evidence. Unless otherwise stated, we make the following choice: $\mu_H = 0.0225$, $\mu_L = 0.015$, $\sigma_D = 0.04$ on an annual basis. Time steps are assumed to be weeks, i.e. $\Delta t = 1/50$. $p_\mu = 0.02$, i.e. the growth rate can be expected to change once a year. The subjective discount factor is $\beta = 0.999$, i.e. 0.95 per year and the risk aversion parameter is set to $\gamma = 3$.[23] The dividend process is

[23] See e.g. Cochrane (2005), p. 455f for a discussion.

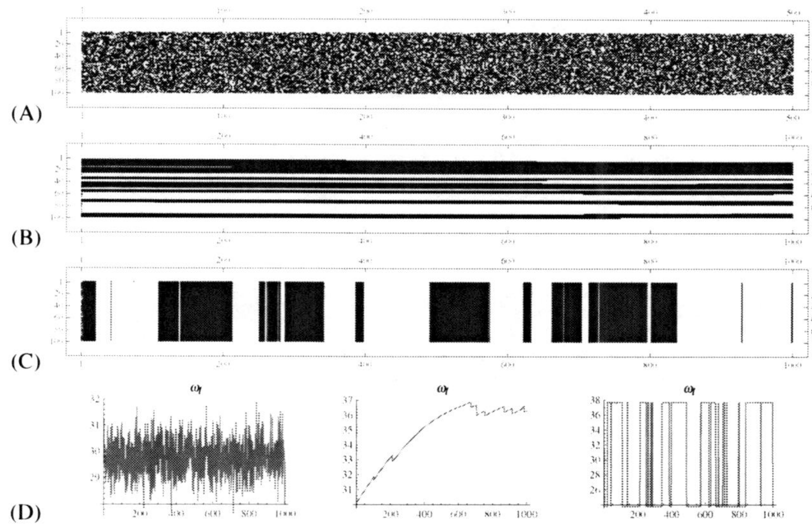

Fig. 8.1 Simulation results for pure internal, external local, and external global effects

initially normalized to $\log D_0 = 1$. Figure 8.1 shows results for $N = 100$ agents and $T = 1{,}000$ time steps. Panels (A) to (C) display the temporal change in the states of agents, where agents are vertically aligned and time steps are on the horizontal axis. Black cells correspond to an agent being in state 1, which means he expects the economy to be in the high growth regime. White cells correspond to state 0, i.e. the agent expects μ_L. Note that in order to obtain the results of all three panels, we hold constant the underlying randomness by initializing the random number generator at the same seed value. In panel (A), $\phi_g \rightarrow \infty$, i.e. only internal, idiosyncratic effects matter, and it can be seen that no local interaction occurs. Results in panels (B) and (C) are due to pure external effects, where in panel (B) only local interaction matters, i.e. $\phi_g \rightarrow 0$ and $\phi_{\cdot\,l} = 1$, whereas in panel (C) global effects dominate, i.e. $\phi_g \rightarrow 0$ and $\phi_{\cdot\,l} = 0$. In the case of pure local effects, clusters occur that are stable over time. On the other hand, with pure global effects, the states of the agents are synchronized through the information available to all of them, and so all agents change states when the information indicates a regime-switch. Panel (D) in the last row contains the price-dividend ratio for the three simulations. For the pure internal case, ω_t fluctuates around a constant value which is given as the mixture of the price-dividend ratio in the high and the low growth regime. For the case of pure local interaction, we observe an increasing behavior of ω_t over time, which is due to the fact that clusters of state 0 agents gradually increase in size. Finally, for pure global effects, i.e. agents are fully synchronized through information, the price-dividend ratio is piece-wise constant, jumping between $\omega_t(\mu_H)$ and $\omega_t(\mu_L)$.

Time series properties of prices The results of pure internal, external local or external global effects in isolation as shown above are obviously the extreme limiting cases of our economy, and it is more meaningful to consider combinations

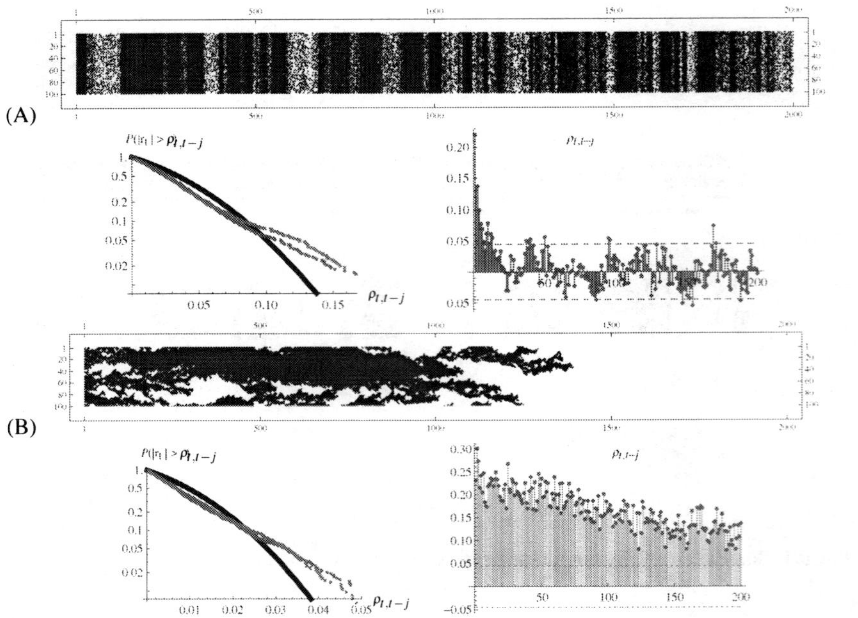

Fig. 8.2 Simulation results for combinations of internal, external local, and external global effects

thereof. In this section, we show that the model is capable of reproducing some important stylized facts of financial time series. In Fig. 8.2, we document results from the combination between internal and external global (panel (A)) and internal and external local (panel (B)) effects. Note that, as above, both simulations are obtained with the same seed value. From panel (A),[24] we observe alternating periods where state 0 and 1 predominates, which corresponds to the information from the filtered signals. However, due to the internal effect, transitions are noisy, i.e. even when the signal suggests that the economy is in the high-growth state, some agents still remain in state 0, and vice versa. Loosely speaking, when markets are bullish, some agents remain pessimistic. In the two diagrams below, we report results for two important statistical tools to analyze the corresponding time series for market prices. The left panel shows the log-log plot of cumulative absolute returns, and the right panel shows a correlogram, i.e. the autocorrelation function up to lag 200 for the absolute returns. The log-log plot for $P(|r_t| > x)$ is frequently used to provide evidence for a scaling or power-law behavior of price movements.[25] Due to space constraints, a comprehensive discussion is beyond the scope of this article. From visual inspection, we can observe a clear linear decay, which provides strong evidence for scaling behavior. The reason is basically the noisy swings in the aggregate

[24] Parameter choices are: $\phi_g = 1.3$, $\phi_l = 0$.

[25] See e.g. Gopikrishnan et al. (2000) or Lux and Marchesi (1999).

states of agents. The corresponding correlogram reveals that there are significant positive autocorrelations in absolute returns only for the first few (<10) lags.[26]

In panel (B), internal and external local effects interact ($\phi_g = 1.3$, $\phi_{\ell} = 1$). As in the previous section, local interaction generates clusters that are stable over time. However, adding internal effects has the effect that there is more variation in the cluster-size. Interestingly, we find a qualitatively similar scaling behavior in absolute returns as in panel (A), documented by the corresponding log-log plot. However, if we now consider the associated correlogram, a fundamentally different result obtains. We observe significant positive autocorrelations for lags up to and beyond lag 200. A slow, i.e. less than exponential decay in the autocorrelation function has been interpreted as long-memory effect and was discussed and documented for financial time series.[27] In our context, the long-memory effect is clearly due to the persistence of the neighborhood effect combined with internal random behavior.

Although the conclusion would require further statistical analysis, we obtain evidence, that our economy is able to reproduce some important stylized facts of market price time series, notably scaling behavior and long-memory in autocorrelations of absolute returns. In particular, our results suggest that both, global and local effects, can explain scaling behavior, but that only local interaction can explain the long-memory in autocorrelations.

The impact of information accuracy In the previous two sections, our focus was on the impact of ϕ on the time series properties of the price-dividend ratio and absolute returns. Thereby, we took the precision of the information, i.e. σ_e as constant, and treated global and local effects in isolation. In this section we consider the impact of a combination thereof, and show how results change with the accuracy of the signal.

First, we will focus on the autocorrelation function of the (absolute) returns series for different signal precisions (σ_e). In Fig. 8.3, we report simulation results for the case where the state of agents is influenced by neighborhood effects as well as by the economy-wide signal and internal effects ($\phi_g = 1$, $\phi_{\ell} = 0.9$).

The left panel displays the autocorrelation function up to lag 50 for absolute (log) returns of the price series with $T = 5,000$ (and as above $N = 100$) for four different values of σ_e, i.e. $\sigma_e \in \{0.001, 0.05, 0.1, 0.15\}$, indicated (in increasing order) by open circles, filled circles, open squares and filled squares. The right panel shows the same for the (log) returns series.

There are basically two important observations. First, the autocorrelation function declines faster than in the case of pure local effects. Second, and more importantly, if the signal is sufficiently precise, we observe significant positive autocorrelation approximately up to lag 20 for absolute returns, and up to lag 10 for returns.

---

[26] Dashed lines indicate the approximate 5% confidence bounds.

[27] There is an extensive literature on long-memory processes and related econometric issues. For an early overview see e.g. Baillie (1996). Evidence is also provided in e.g. Gopikrishnan et al. (2000).

(A)
(B)
(C)

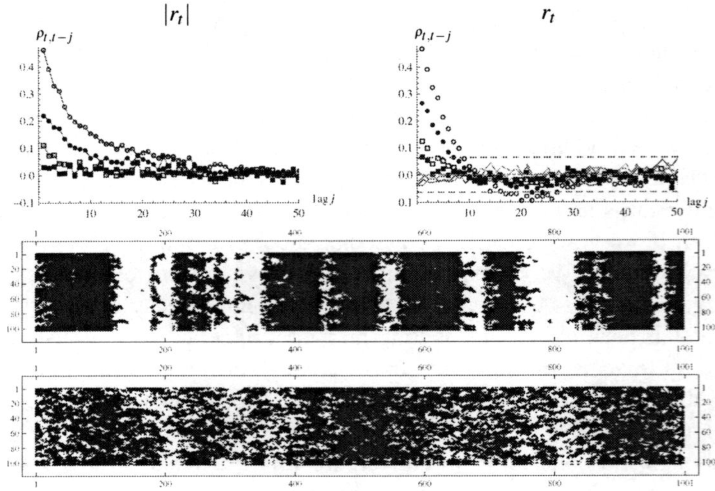

Fig. 8.3 The autocorrelation function for different levels of the signal precision

However, as the precision of the signal declines, the autocorrelations decline and eventually become insignificant.

If σ_e is small, agents get a strong signal about the regime and thus a large fraction of all agents is in the same state. However, due to local effects, there are still small clusters of agents remaining in the previous state, although new information is available. (This is illustrated in panel (B) of Fig. 8.3). The size of these clusters declines over the subsequent periods, which can explain the positive autocorrelations. On the other hand, if σ_e increases, i.e. the accuracy of the signal declines, the coordinating mechanism of the global signal is weak which implies that the distribution of states over time is more homogenous (see panel (C) of Fig. 8.3). This in turn means that price changes are only weakly correlated over time. Note, that the global information (although being imprecise) prevents the economy from being eventually dominated by either optimists or pessimists as this is the case for the isolated local interaction mechanism (see panel (B) in Fig. 8.2). Taken together, the results suggest that in our artificial economy we find that an increase in the precision about the regime of the economy may lead to a higher predictability of returns, which may be interpreted as a weaker degree of market efficiency.

As a final point, we consider the impact of the information accuracy in our economy on some proxy for the risk premium. As mentioned in the introduction, Veronesi (2000) f.ex. analyzes the impact of information precision in the context of a pure-exchange economy and finds that the risk premium *increases* with information accuracy. The reason for this somewhat surprising result is that in his consumption-based asset pricing framework, the risky asset is a hedge against future consumption. With more precise signals, the covariation increases and thus the risky asset demands a higher risk premium.

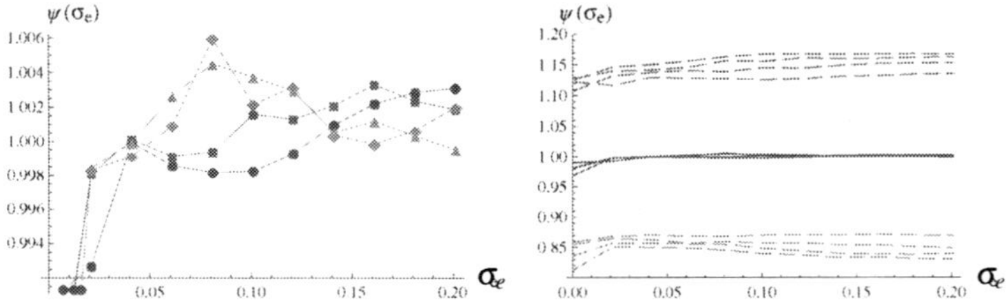

Fig. 8.4 The ratio of covariation as function of the signal precision

Therefore, we simulate our economy for different signal precisions and calculate the ratio

$$\psi(\sigma_e) = \frac{Cov(\log(P_t/P_{t-1}); \log(D_t/D_{t-1}))}{Var(\log(D_t/D_{t-1}))},$$

i.e. the covariance between price and dividend changes to the variance of the log dividend changes. Figure 8.4 reports the corresponding results for $\phi_g = 1.5$, $\phi_l \in \{0.8, 0.85, 0.9, 0.95\}$.[28] Although it would be tempting to conclude from the left panel, that the covariation ratio increases as the signal precision declines, which would provide evidence against the results of Veronesi (2000), the effect is actually very small and is statistically not significant. This is obvious from the right panel, where we additionally report the one standard deviation bounds. Thus, on the basis of these results, we do not find some significant effect of local and global interaction on the risk premium in our economy.

8.4 Conclusion

In this article, we consider a standard pure-exchange economy where the endowment process follows a simple hidden Markov chain and agents have incomplete information about the regime. We extend this framework by considering local interaction between a finite number of risk-averse agents, thereby dropping the representative agent assumption.

Although results are largely tentative, our economy displays interesting effects. The interplay of internal, external local and external global influences can explain different temporal patterns of the price-dividend ratio and is able to produce stylized facts such as scaling behavior and long memory in autocorrelations. It has been argued that only local interaction effects can account for a slow decay of the autocorrelation function. Furthermore, by varying the precision of the economy-wide signal, we find that the predictability of price changes is inversely related to the accuracy of the signal. This provides some evidence that there is an inverse relation

[28] $N = 100$, $T = 1{,}000$. We repeated the simulation for every value of σ_e and ϕ_l with the same initial random seed value, and averaged over $K = 300$ different seed values. On a pentium 4, 3.0 GHz this required a runtime of approximately 280.000 s.

between signal precision and market efficiency. With respect to the risk premium, our results suggest that the impact of information precision is at best marginal.

If these results can be supported by a formal analysis awaits further research.

Acknowledgements We thank two anonymous referees for helpful and valuable comments.

References

K. Back. Incomplete and asymmetric information in asset pricing theory. In M. Frittelli and W. Runggaldier, editors, *Stochastic Methods in Finance*, Springer Lecture Notes in Mathematics 856. Springer, 2004.

R. T. Baillie. Long memory processes and fractional integration in econometrics. *Journal of Econometrics*, 73:5–59, 1996.

S. Bikhchandani, D. Hirshleifer, and I. Welch. A theory of fads, fashion, custom, and cultural change as informational cascades. *Journal of Political Economy*, 100(5):992–1026, 1992.

M. Brennan and Y. Xia. Stock price volatility and equity premium. *Journal of Monetary Economics*, 47:249–283, 2001.

J. H. Cochrane. *Asset Pricing*. Princeton University Press, Princeton, 2nd edition, 2005.

R. Cont and J.-P. Bouchaud. Herd behavior and aggregate fluctuations in financial markets. *Macroeconomic Dynamics*, 4(2), 2000.

J. B. Detemple. Asset pricing in a production economy with incomplete information. *Journal of Finance*, XLI(2):383–391, 1986.

M. U. Dothan and D. Feldman. Equilibrium interest rates and multiperiod bonds in a partially observable economy. *Journal of Finance*, XLI(2):369–382, 1986.

J. D. Farmer and S. Joshi. The price dynamics of common trading strategies. *Journal of Economic Behavior & Organization*, 49(2):149–171, 2002.

G. Gennotte. Optimal portfolio choice under incomplete information. *Journal of Finance*, XLI(3): 733–749, 1986.

P. Gopikrishnan, V. Plerou, Y. Liu, N. L. A. Amaral, X. Gabaix, and H. E. Stanley. Scaling and correlation in financial time series. *Physica A*, 287:362–373, 2000.

J. D. Hamilton. Regime-switching models. In S. N. Durlauf and L. E. Blume, editors, *The New Palgrave Dictionary of Economics*. Palgrave Macmillan, 2008.

D. Hirshleifer and T. Shumway. Good day sunshine: Stock returns and the weather. *Journal of Finance*, 58(3):1009–1032, 2003.

C. H. Hommes. Interacting agents in finance. In L. E. Blume and S. N. Durlauf, editors, *New Palgrave Dictionary of Economics*. Palgrave Macmillan, 2nd edition, 2006.

A. Kirman. Whom or what does the representative individual represent? *Journal of Economic Perspectives*, 6(2):117–136, 1992.

B. LeBaron. Agent-based computational finance. In K. L. Judd and L. Tesfatsion, editors, *Agent-based computational economics*, volume 2 of *The Handbook of Computational Economics*. North-Holland, Amsterdam, 2006.

R. E. Jr. Lucas. Asset prices in an exchange economy. *Econometrica*, 46(6):1429–1445, 1978.

T. Lux and M. Marchesi. Scaling and criticality in a stochastic multi-agent model of a financial market. *Nature*, 397:498–500, 1999.

R. Mehra and E. C. Prescott. The equity premium in retrospect. In G. M. Constantinides, M. Harris, and R. M. Stulz, editors, *Handbook of the Economics of Finance: Financial Markets and Asset Pricing*, volume 1B of *Handbooks in Economics*. North-Holland, Amsterdam, 2003.

R. Mehra and E. C. Prescott. The equity premium: A puzzle. *Journal of Monetary Economics*, 15: 145–161, 1985.

S. Morris. Contagion. *Review of Economic Studies*, 67:57–78, 2000.

A. Orlean. Bayesian interactions and collective dynamics of opinion: Herd behavior and mimetic contagion. *Journal of Economic Behavior & Organization*, 28:257–274, 1995.

E. M. Jr. Saunders. Stock prices and wall street weather. *American Economic Review*, 83:1337–1345, 1993.

D. Stauffer and D. Sornette. Self-organized percolation model for stock market fluctuations. *Physica A*, 271:496–506, 1999.

P. Veronesi. How does information quality affect stock returns? *Journal of Finance*, LV(2):807–837, 2000.

P. Veronesi. Stock market overreaction to bad news in good times: A rational expectations equilibrium model. *The Review of Financial Studies*, 12(5):975–1007, 1999.

N. J. Vriend. Ace models of endogenous interactions. In K. L. Judd and L. Tesfatsion, editors, *Agent-based computational economics*, volume 2 of *The Handbook of Computational Economics*. North-Holland, Amsterdam, 2006.

A. W. Wilhite. Economic activity on fixed networks. In K. L. Judd and L. Tesfatsion, editors, *Agent-based computational economics*, volume 2 of *The Handbook of Computational Economics*. North-Holland, Amsterdam, 2006.

S. Wolfram. *A new kind of science*. Wolfram Media, Champaign, 2002.

Chapter 9
Long-Term Orientation in Trade

Gert Jan Hofstede, Catholijn M. Jonker, and Tim Verwaart

Abstract Trust does not work in the same way across cultures. This paper presents an agent model of behavior in trade across Hofstedes cultural dimension of long-term vs. short-term orientation. The situation is based on a gaming simulation, the Trust and Tracing game. The paper investigates the micro-dynamics of trust in a trade relationship, focusing on one dimension of culture only. This is a deliberate simplification that is made in order to highlight the role of this particular dimension of culture. The dimension is relevant for trade between East Asia and the Western world. The paper discusses the cultural dimension of long-tern vs. short-term orientation and specifies decision rules for trading agents, differentiated across the dimension. The paper contributes to bridging the gap between macro level analysis of intercultural trade and micro level knowledge about culturally determined behaviour. To this end the agent model is implemented in a multi-agent simulation. Potential applications of the model are to be found primarily in research into the formation and performance of supply chains and networks, and secondarily in training, education, and advice systems. The implementation of the model has been verified to qualitatively represent the effects expected on the basis of Hofstedes theory.

G.J. Hofstede
Wageningen University, Hollandseweg 1, 6706 KN Wageningen, the Netherlands, e-mail: gertjan.hofstede@wur.nl

C.M. Jonker
Delft University of Technology, Mekelweg 4, 2628 CD Delft, the Netherlands, e-mail: C.M.Jonker@tudelft.nl

T. Verwaart
LEI Wageningen UR, Postbus 29703, 2502 LS Den Haag, the Netherlands, e-mail: tim.verwaart@wur.nl

9.1 Introduction

Trade requires trust. A buyer has to trust the seller, or the virtues of the product, or if he trusts neither he could still trust the enforcement mechanisms that can be put into operation in case of cheating. For a seller it is the same, except that he usually knows about the product, so that he is on the knowing side of an information asymmetry.

Trust does not work in the same way across cultures (Hofstede, 2007). For instance, in some societies the emphasis is on interpersonal trust based on kinship; in others it is on interpersonal trust based on other markers; yet in others it is in impersonal institutions such as certification or law. Gorobets and Nooteboom (2006), demonstrated on the basis of a multi-agent simulation that economic systems based on trust as well as systems based on opportunism may be viable in different societies.

This paper investigates the micro-dynamics of trust in a trade relationship. It does so by specifying an agent model for trade in a product with a hidden quality attribute. The situation is based on a simulation game, the Trust and Tracing game (Meijer et al., 2006; Jonker et al., 2006). The focus of study of this game is on trust in a business partner when acquiring or selling commodities with invisible quality.

When playing the game with people from different parts of the world, different patterns of trade and trust emerge (Meijer et al., 2006). A multi-agent model of the interactions in the game may help to improve the understanding of effects of culture on trade, and may help to understand if different economic systems and institutions may have been differently efficient in different parts of the world.

This paper focuses on one aspect of culture only. This is a deliberate simplification that is made in order to highlight the role of this particular dimension of culture. The 5-dimension model by Hofstede (2001) allows for this approach. The paper first discusses the cultural dimension of long-term vs. short-term orientation. It then specifies the decision rules for the trading agents. After this, results of test runs are presented. Some conclusions and caveats round off the paper.

9.2 Long- vs. Short-Term Orientation

Behaving as a good, upstanding member of the group is at the core of the lives of all beings that live in social groups (Wilson, 2007). Human beings are intensely social and they exemplify the point very well. Ensuring the successful functioning of our social groups is a basic requirement for survival. We spend up to twenty years being taught how to act as virtuous members of society. But how to be virtuous? It turns out that different societies have found different answers to that question. In some, rationality is a prominent virtue; in others, common sense. In some, virtue consists primarily in honouring tradition; in others, it consists more of becoming prosperous. The issue of whether the sources of virtue are to be found in the past and present on the one hand, or in the future on the other hand, is one of the basic dimensions of culture found by Hofstede (2001). It is a dimension not easily recognized by Western minds. The other four dimensions (individualism / collectivism, power distance, masculinity / femininity, uncertainty avoidance) in Hofstedes model are more often mentioned in research. Yet this fifth dimension, called "long-term vs. short-term orientation" (LTO) by Hofstede, is very important for trade. If nothing else, its strong correlation with economic growth across 39 countries in the period 1970–2000 testifies to this (Hofstede and Hofstede, 2005, p. 223).

Hofstede (2001, p. 359) gives the following definition of the LTO-dimension:

> Long Term Orientation stands for the fostering of virtues oriented towards future rewards, in particular, perseverance and thrift. Its opposite pole, Short Term Orientation, stands for the fostering of virtues related to the past and the present, in particular, respect for tradition, preservation of "face" and fulfilling social obligations.

Both positively and negatively rated values of this dimension are found in the teachings of Confucius, but the dimension also applies to countries without a Confucian heritage, for instance Brazil. Long-term orientation needs not be correlated with collectivism, as it is in the very collectivist Southeast Asian countries. Within Europe, Finland and the Netherlands are fairly long-term oriented, whereas Sweden and Germany are more short-term oriented. Table 9.1 displays distinctions that are relevant to the work in this paper.

The LTO-dimension can be predicted to have some effect on behaviour of participants in the Trust And Tracing game. The game has a limited time frame, so a long term oriented strategy may not be a successful one to win the game. The games reward system is about gains made during the game run, not about building a strong market position; the game is finite. One typical characteristic of long-term oriented thinking is that it is not limited to the time frame or to the logical puzzle set by any single event. Chinese, Japanese, Korean and Vietnamese culture are long-term oriented. Hampden-Turner and Trompenaars (1997) were thinking about this when they entitled their book about Asian business values Mastering the infinite game. Asian logic is not atomistic.

Hofstede (2001) performs an extensive literature review that confirms the importance of long-term orientation to trade. Long-term orientation opposes the disruption of harmonious relationships. This does not mean that conflict or competition are nonexistent. It means that they will not be framed as such. There are strong encouragements to self-discipline, and any emotion that might cause repercussions is

Table 9.1 A summary of relevant distinctions between norms in long term oriented and short term oriented societies (source: Hofstede, 2001, p. 367)

Short term oriented	Long term oriented
Immediate gratification of needs expected	Deferred gratification of needs accepted
Traditions are sacrosanct	Traditions adaptable to changed circumstances
Short-term virtues taught: social consumption	Long-term virtues taught: frugality, perseverance
Spending	Saving, investing
The bottom line	Building a strong market position
Analytical thinking	Synthetic thinking

repressed. Subordinates are very ready to comply, because it is in the interest of the social entity as a whole. Usefulness to the whole of society is an important criterion for behaviour, and it takes precedence over truth. Making statements that are not factual in order to achieve benefits for the group is not a sin, but a virtue.

Long-term orientation is correlated with self-effacement. This was found by Minkov (2007) in a meta-analysis of data from the World Values Survey. The individual thinks of itself as a small element within the continuity of life. Hence, learning and developing ones capacities is more important than winning a particular game, or obtaining a particular result. In contrast, short-term orientation correlates with self-enhancing values in which being successful in a game is a desirable thing that will improve ones reputation.

It can thus be expected that the performance of long-term oriented traders is much dependent upon factors that are not within the scope of the experimental setting. They will be more aware of possible adverse real-world consequences of what they do during the game than will participants from short-term oriented cultures. Of course this problem needs not affect the performance of agents – but it will affect validation against the behaviour of human traders.

9.3 The Effect of LTO on Trade Processes

Figure 9.1 presents a process model of a trading agent, inspired by the setting of the Trust and Tracing game. The focus of this game is on trust in a business partner when acquiring or selling commodities with invisible quality. There are five roles: traders (producers, middlemen and retailers), consumers and a tracing agency. Typically there are four producers, four middlemen, four retailers and eight consumers, to reflect the multiple steps and oligopoly character of most supply networks. The real quality of a commodity is known by producers only. Sellers may deceive buyers with respect to quality, to gain profits. Buyers have either to rely on information provided by sellers (Trust) or to request a formal quality assessment at the Tracing Agency (Trace). This costs a tracing fee for the buyer if the product is what the seller stated (honest). The agency will punish untruthful sellers by a fine. Results

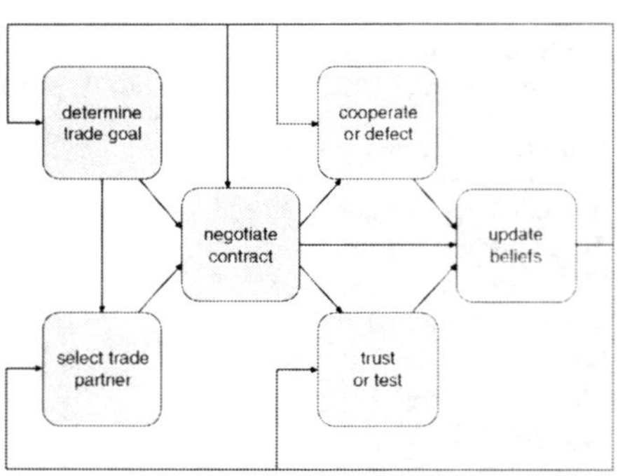

Fig. 9.1 Processes and internal information flow of a trading agent

of tracing are reported to the requestor only or by public disgrace depending on the game configuration. A strategy to be a truthful seller is to ask for a trace before selling the product. In that case sellers use the tracing report as a quality certificate. Middleman and retailers have an added value for the network by their ability to trace a product cheaper than a consumer can.

For each of the processes represented in Fig. 9.1, the paper elaborates on the expected difference in behaviour between from long- and short-term oriented cultures, based on Hofstedes work.

Partner Selection. Long term oriented traders aim to develop lifetime relations. Not only a transaction's calculated pay-off counts; the resulting relation is of value by itself. Given the choice, a long term oriented trader would not neglect a proposal from a known relation in order to make a one more attractive one time deal with a stranger.

Short term oriented traders are not interested to develop long lasting relations. They want to make attractive deals as soon as possible. Their preferences in partner selection depend on other dimensions of culture, but they may be interested in doing business with high status partners to show off.

Trade Strategy Selection. Long term oriented traders will follow a thrifty strategy. They trade valuable high quality products if it serves their purpose, but they do not need them to show off. They avoid the risk of damaging their relations and do not trade commodities that they are not certain about. Rather will they trade low quality, avoiding the risk of unintentionally selling low quality for high quality, or being forced to show distrust in suppliers by tracing their deliveries. Even if they like to gamble, they will not do it in business. They invest in relations and will not defect.

Short term oriented traders are more opportunistic. They analyze a deal in terms of pay-off (which may include future prospects), rather than a relational network. More than the long term oriented, and depending on other dimensions of culture, they like to show off and are prepared to take risks if this is necessary to save face and meet social demands.

Negotiation Behavior. Long term oriented traders show patience. They do not break off negotiations. They do not overcharge. A first proposal may be modest, but they do not rapidly give in.

Extremely short term traders are impatient. They want rapid deals. If they give in they do it quickly and with substantial concessions. If partners do not make concessions too, they break off easily and try their luck elsewhere.

Truthful or Untruthful Delivery. After an agreement has been reached, it comes to delivery. If the quality of the commodity is invisible at first sight, the supplier can be opportunistic and try to deliver a lower quality product than agreed upon, thus making an extra profit. Given the opportunity (and depending on other preferences), a short term oriented supplier may be tempted do so, after calculating the possible gain. A long term oriented supplier will not put a relation with a customer at stake by deceiving.

Trust or Distrust. Long term oriented traders cannot take the risk of unintentionally reselling low quality products for high quality products. They do not easily trust new relations. Trust has to be earned. They deliver truthfully and demand equal seriousness from their suppliers. Once relations have developed, they may signal their valuation of the relation by showing trust. When they find out that their trust has been abused, they lose all respect and banish the supplier from business.

The short term oriented traders trust or distrust on the basis of calculated risk and preferences, but the scope of their calculations is no more than the bottom line of the current transaction.

Maintenance of Beliefs About Partners. People from long term oriented cultures value their relations. They value a deal not only by the financial pay off, but also by the relational gains. They invest in relations by behaving truthfully and by trusting their partners. They have no respect for others that put their relations at stake for some short term profit. If they turn out to be deceived by a business partner they will not easily forgive the deceiver.

People from short term oriented cultures find it hard to understand the sacrifice of the long term oriented. The short term oriented tend to grab a chance for an easy profit and are willing to put their relations at stake for it (the other dimensions of culture, personality and circumstances determine the extent to which they will actually defect). They can understand that a business partner may be tempted to defect if a profitable opportunity occurs, and they have trouble understanding that people from long term oriented cultures cannot.

9.4 Representation in Agents

This section formalizes the knowledge about the influence of power distance on trade processes, formulated in Sect. 9.2. The relevant attributes of transactions from this viewpoint are the economic value of the transaction, the quality of the traded goods as a status attribute in its own, and a perceived risk that the trade partner will not fulfill its contractual obligations or will not trust the delivery.

The negotiation model uses a utility function to compare bids.

$$U_{bid} = w_1 f_1(value) + w_2 f_2(quality) + w_3 f_3(risk) \tag{9.1}$$

with $w_1 + w_2 + w_3 = 1$, and w_i in $[0,1]$, for all i. $f_1 \ldots f_3$ normalize value, quality, and perceived risk of the transaction in the interval $[0,1]$. Agents are modeled to have personal traits that determine their decisions and behavior. The traits are implemented as real values in $[0,1]$. Some traits are related to negotiation behavior; the most important are the weight factors in the utility function, impatience, concession factor and other parameters in the negotiation architecture of Jonker and Treur (2001). Some traits are related to the delivery decision and trust (honesty, confidence), some to the update functions that maintain beliefs about partners trustworthiness, benevolence, and acceptability as a trade partner. In addition to the traits, each agent has a long term orientation indicator – a real number in $[0,1]$. Traits and LTO-indicator are not visible to other agents. Furthermore, each agent has a label, that represents it societal status. The status label is visible for other agents. In the partner selection process, agents select the most acceptable partner. They try and open new negotiations by replying if a proposal has been received recently or by taking the initiative to propose to a partner. Agents may ignore a proposal if the partner is not sufficiently acceptable. The paper formulates rules for the decisions of the agents, based on the values of personal traits, LTO, and beliefs about the partner. An example of such a rule is given below. It uses a uniform random variable and increased probability (up to 0.5) to stop for the reason gap to the extent the agent is impatient, but impatience is tempered if the agent stems from LT-oriented culture or deals with a high-status agent.

Rule 1. An STO agent tends to follow a *quality-minded* strategy, represented by relatively high values of w_2 and w_3 in (9.1); an LTO agent tends to follow a *thrifty* strategy: relatively low value of w_2. The denominator N guarantees that the sum of the weight factors equals 1.

```
if cultural_script_contains(long-term-orientation (L: Real))
    and agent_trait_value(value_preference, P: Real)
    and agent_trait_value(quality_preference, Q: Real)
    and agent_trait_value(risk_aversion, R: Real)
    and N: Real = P + Q*(1 - L)*2 + R*min(1, (1 - L)*2)
then weight_for_value (P / N)
    and weight_for_quality (Q*(1-L)*2 / N)
    and weight_for_risk (R*min(1, (1-L)*2) / N);
```

Rule 2. In negotiations, an STO agent shows impatience if partner makes unrealistic bids or makes little progress, unless the partners societal status is high. The following rules express how the probability that a negotiation will be terminated if $U < M$ or $N < M$, depends on impatience, LTO, and status difference. Furthermore the rules record the reason for termination: there is either a gap between the partners proposal and what the agent finds reasonable, or the partner shows too little intention to accommodate the agent.

```
if cultural_script_contains (long-term-orientation(L: Real))
and current_round(X: Integer)
and current_negotiation (T: Trader, X, C: Commodity_list)
and others_bid_utility_in_round(U: Real, X)
and agent_trait_value(cut_off_value, M: Real)
and U < M
and agent_trait_value(impatience, I: Real)
and agent_label(status, S: Real)
and partner_model_contains_belief (T, status, Y: Real)
and random(0, 1, Z: Real)
and I * (1 - max(L, (1 - L)*(Y - S)))*0.5 > Z
then stop_negotiation(T, X, C, gap);

if cultural_script_contains (long-term-orientation(L: Real))
and current_round(X: Integer)
and current_negotiation (T: Trader, X, C: Commodity_list)
and progress_in_bids(X-3, X, N: Real)
and agent_trait_value(minimal_progress, M: Real)
and N < M
and agent_trait_value(impatience, I: Real)
```

```
and agent_label(status, S: Real)
and partner_model_contains_belief (T, status, Y: Real)
and random(0, 1, Z: Real)
and I * (1 - max(L, (1 - L)*(Y - S)))*0.5 > Z
then stop_negotiation(T, X, C, no_accom);
```

Rule 3. An STO agent delivers opportunistically; an LTO agent has a high threshold to defect deliberately. The threshold is used in the decision to cooperate or defect. The probability that an agent will cooperate is a monotonous function of the threshold, but also depends on other factors, like beliefs about the other agent and the relation.

```
if cultural_script_contains(long-term-orientation (L: Real))
and agent_trait_value(honesty, H: Real)
then deceit_treshold(H + (1 - H)*L);
```

Rule 4. An LTO agent exercises restraint to trace, cautious to preserve the relation; an STO agent only exercises restraint if partners status is high.

```
if cultural_script_contains(long-term-orientation (L: Real))
and current_round (X: Integer)
and deal_in_round (T: Trader, B: Bid, X)
and partner_model_contains_belief(T , trustworthiness, W: Real)
and agent_label(status, S: Real)
and partner_model_contains_belief(T , status, Y: Real)
and random(0, 1, Z: Real)
and W + (1-W)*max(L, (1 - L)*(Y - S)) < Z
then to-be-traced(B);
```

Rule 5. An LTO agents trust is more deeply affected than an STO agents if a partner defected. Beliefs about another agents traits and attitudes are updated on the basis of an experience value E, for a positive experience as $B_{t+1} = (1-d^+)B_t + d^+E$, and for a negative experience as $B_{t+1} = (1-d^-)B_t + d^-E$, with $d^- \geq d^+$ and the endowment factor $e = d^+/d^-$.

```
if cultural_script_contains(long-term-orientation (L: Real))
and agent_trait_value(base_neg_update_factor, N: Real)
and agent_trait_value(base_endowment_factor, E: Real)
and lto_endowment_factor (F: Real)
then neg_update_factor(N*(1 - L) + L)
and endowment_factor(E*(1 - L) + F*L);
```

Rule 6. An STO agent has stronger preference to select high-status partners than an LTO agent. The acceptability of a partner depends primarily on its believed fairness, but an STO agent also takes the partners societal status into account. It likes to show off, while an LTO agent is interested to have long-standing business relationships, independent of partners status.

```
if cultural_script_contains(long-term-orientation (L: Real))
and partner_model_contains_belief(T: Trader, fairness, F: Real)
and agent_label(status, S: Real)
and partner_model_contains_belief(T , status, Y: Real)
then acceptability (T, F + (1 - F)*(1 - L)*max(0, Y - S));
```

Rule 7. An STO agent has aversion against partners that did not respect it (rule 2) or simply did not reply to a bid in an ongoing negotiation. It drastically reduces its fairness belief about the partner if a partner did not show respect by making acceptable proposals.

```
if cultural_script_contains(long-term-orientation (L: Real))
and current_round(X: integer)
and stop_negotiation(T: Trader, X, C: Commodity_list, gap)
and partner_model_contains_belief(T: Trader, fairness, F: Real)
and neg_update_factor(N: Real)
then new_fairness(T, (1 - max(N, 1  L))*F);

if cultural_script_contains(long-term-orientation (L: Real))
and current_round(X: integer)
and stop_negotiation(T: Trader, X, C: Commodity_list, no_accom)
and partner_model_contains_belief(T: Trader, fairness, F: Real)
and neg_update_factor(N: Real)
then new_fairness(T, (1 - max(N, 1 - L))*F);

if cultural_script_contains(long-term-orientation (L: Real))
and current_round(X: integer)
and stop_negotiation(T: Trader, X, C: Commodity_list, no_reply)
and partner_model_contains_belief(T: Trader, fairness, F: Real)
and neg_update_factor(N: Real)
then new_fairness(T, (1 - max(N, 1 - L))*F);
```

9.5 Experimental Verification

The implementation of the rules developed in the previous section was verified at two levels. First the rules were verified at the level of individual decisions in scenarios of one-to-one agent interactions. Secondly the rules were verified in multi-agent simulations. In the multi-agent simulations, eight supplier agents and eight customer agents were trading commodities with a varying quality, visible to the supplier but invisible to the customer.

To assess the sentisitivity of the model for parameter settings, a thousand runs of 100 time steps were performed to test the sensitivity for a set of 10 parameters. For each parameter a random value was independently drawn from the uniform distribution for each run. Eight supplier agents and eight customer agents were all configured homogeneously with the parameters drawn for the run. The number of successful transactions was observed per run. The observed output appeared to be particularly sensitive for three parameters: the LTO-index, the weight factor w_2 for quality in the utility function, and the initial fairness belief about other agents. A linear model of the number of successful transaction was fitted for the ten parameters. The adjusted R^2 value was 69.0%.

Table 9.2 presents estimated values of the bottom marginal variance and the top marginal variance associated with the parameters for this model. The bottom marginal variance is the percentage of total variance of the output that is no longer

Table 9.2 Interval used for generation of random values, and bottom marginal variance (BMV) and top marginal variance (TMV) of 10 parameters used for sensitivity analysis

Parameter	Interval	BMV	TMV
LTO	[0.01, 0.99]	27.1	25.6
Status_difference	[0.01, 0.99]	0.4	0.3
Honesty	[0.01, 0.99]	4.3	4.2
Initial_fairness	[0.01, 0.99]	6.5	7.8
Impatience	[0.01, 0.99]	2.6	4.0
Endowment_coefficient	[0.50, 0.99]	0.4	0.3
Lto_endowment_coefficient	[0.20, 0.50]	0.0	0.0
Weight_Q	[0.01, 0.99]	26.3	27.1
Weight_R	[0.01, 0.99]	0.2	0.2
Negative_update_factor	[0.01, 0.99]	0.8	0.0

explained if the parameter is removed from the model; the top marginal value is the percentage of variance explained by a simple regression model of the parameter, as explained in Jansen et al. (1994).

The percentage of variance explained is rather low. Adding an interaction term for the three dominant parameters resulted in an R^2 value of 74.3%. A better fit could not be found. The remaining variance may be attributed to random effects in the simulation. In order to assess the variance introduced by random effects, 30 random parameters sets were generated. With each of these sets, 33 runs of 100 time steps were performed. For each set of 33 run with equal parameters, the standard deviation of the number of successful transactions was computed. The results for the standard deviation as percentage of the mean value are: minimum value 6.5, 25-percentile 8.2, median 9.6, mean 10.6, 75-percentile 12.9, maximum 17.9.

Scenarios were run to test the effect of the LTO index in different homogeneous and heterogeneous configurations, in 6 sets of 100 runs. Hundred independent parameter sets were generated at random, except for LTO and for status. The random parameter sets were reused for each set of 100 runs, in order to enable pairwise comparison of individual runs. Values of LTO were fixed to LTO = 0.2 for all agents in the first 100 runs, LTO = 0.8 for all agents in the second 100 runs, LTO = 0.2 for suppliers and LTO = 0.8 for customers in the third 100 runs, vice versa in the fourth 100 runs. In the fifth 100 runs, 4 suppliers and 4 customers were given a value of LTO = 0.2, the other agents were given LTO = 0.8. Status was equal for all agents in the first 500 runs. In the last set of 100 runs, status was assigned at random at each individual agent. Table 9.3 presents average results of multi-agent simulations.

The high frequency of tracing and defection and the high level of average quality where STO-agents are involved is as expected and corresponds to the rules. The differences in number of transactions are more puzzling. All differences are significant at the two-sided 99% level, except the difference between run sets 1 and 3 and run sets 2 and 4. The high volumes of transactions occur in particular in individual runs with STO-agents where the w_2 parameter is very high. This makes the agents careless about money. For realistic simulations this parameter should not be

Table 9.3 Average statistics of sample runs with long-term oriented (LTO = 0.8) and short-term oriented (LTO = 0.2) agents; the presented figures are mean values of 100 runs, each having a duration of 100 time steps; eight suppliers and eight customers can select each other for negotiation, exchange bids, deliver truthfully or defect, and request a trace

Supplier culture Customer culture	STO STO	LTO LTO	STO LTO	LTO STO	mixed mixed	status status
Number of successful transactions	91	43	92	42	69	57
Number of traces requested	11.8	2.2	3.3	9.1	6.9	6.0
Number of defections revealed by tracing	0.7	0.0	0.2	0.1	0.3	0.4
Number of unrevealed defections	2.4	0.5	3.3	0.3	1.8	1.8
Average quality agreed in transactions	0.9	0.7	0.9	0.8	0.9	0.9

set greater than 0.5. Another option might be to replace the additive utility function by a Cobb–Douglas type function. However, the latter requires a recalibration of the applied negotiation architecture, for which realistic parameter values were assessed for use with the additive utility function in human experiments.

A second factor that strongly influences the results is the initial setting of the fairness belief that agents maintain about other agents. It plays an important role in the partner selection process. If its value is low, the experience of a single successful transaction may improve the mutual beliefs to such extent that the agents stick together forever. So the initial setting of this parameter should be a realistic value, e.g., equal to the minimum-utility that agents apply in negotiations.

The results verify the implementation of the rules in the agents, in the sense that they behave and interact as expected. However, these tests do not validate the model to represent real trade processes. They only verify the correct implementation of the rules.

9.6 Conclusion

Hofstede and numerous other authors (e.g. Hampden-Turner and Trompenaars, 1997) describe cultural differences and their effect on trade at the individual level. The present paper contributes to bridging the gap between those approaches with agent-based modelling. In agent-based economics, individual traders are modeled as intelligent agents cooperating in an artificial trade environment. The agents are modeled as closely as possible to authentic human behavior. In recent papers the differences between such agents is no longer solely attributed to differences in their individual economic situations. Aspects such as personality and attitude are considered as well, see for example Jager and Mosler (2007). Without considering such aspects, the simulations will not correspond to reality. With respect to formalizing the important influence of cultural background on trade, we only found a few papers. The papers study trade at the macro-level. For instance, Kónya (2006) presents an equilibrium analysis on the amount countries invest in learning another language and culture and the size and well fare of those countries. Bala and Long (2005) present a formal model of the influence of trade on culture, i.e., the reverse direction of influence as studied in the current paper. Other literature also uses macro-level models, such as the gravity model to study the correlation between culture and trade, e.g., Guo (2004).

The contribution of the present paper is the formalization and simulation of culture with respect to the influence of Hofstedes cultural dimension of long-term vs. short-term orientation. This dimension is relevant for trade between East Asia and the Western world. The formalization and simulation have been carried out on the micro-level, i.e., on the level of the individuals participating in trade. The trader behavior is formalized in the form of rules that take long-term vs. short-term orientation of the parties involved into account. The agents reason with a perceived model of the parties they consider for trading. These perceived models do not contain estimates of the culture of the other parties.

The implementation of the model has been verified to qualitatively represent the effects expected on the basis of Hofstedes theory, if agents were configured either extremely long-term oriented or extremely short-term oriented. However, validation against empiric data from experiments are required to calibrate parameters to actual behavior and to scale Hofstedes LTO index realistically to the simulation parameter setting. Validation and refining and tuning the model to real-life situations remains for future work.

The model has been developed as a research tool to experiment with intercultural trade processes in different institutional settings. Other potential application areas are training programs for business schools and international companies.

Acknowledgements The authors thank John Wolters for engineering the simulation, the reviewers for their comments, and Florian Hauser for his help with LaTeX.

References

V. Bala and N. V. Long. International trade and cultural diversity with preference selection. *European Journal of Political Economy*, 21:143–162, 2005.

A. Gorobets and B. Nooteboom. Agent based modeling of trust between firms in markets. In C. Bruun, editor, *Advances in Artificial Economics*, LNEMS 584, pages 121–132. Springer, Berlin, 2006.

R. Guo. How culture influences foreign trade: evidence from the U.S. and China. *Journal of Socio-Economics*, 33:785–812, 2004.

C. Hampden-Turner and F. Trompenaars. *Mastering the Infinite Game: How East Asian Values are Transforming Business Practices*. Capstone, Mankato, MN, 1997.

G. Hofstede. *Cultures Consequences, Second Edn.* Sage Publications, Thousand Oaks, CA, 2001.

G. Hofstede and G. J. Hofstede. *Cultures and Organizations: Software of the Mind, Third Millennium Edn.* McGraw-Hill, London, 2005.

G. J. Hofstede. Trust and transparency in supply netchains: a contradiction? In W. Y. C. Wang et al., editor, *Supply Chain Management: Issues in the New Era of Collaboration and Competition*, pages 105–126. Idea Group, 2007.

W. Jager and H. J. Mosler. Simulating human behavior for understanding and managing environmental dilemmas. *Journal of Social Issues*, 63:97–116, 2007.

M. J. W. Jansen, W. A. H. Rossing, and R. A. Damen. Monte carlo estimation of uncertainty contributions from several independent multivariate sources. In J. Grasman and G. van Straten, editors, *Predictability and nonlinear modelling in natural sciences and economics*, pages 334–343. Kluwer, Dordecht, 1994.

C. M. Jonker and J. Treur. An agent architecture for multi-attribute negotiation. In N. Nebel, editor, *Proceedings of the 17th International Joint Conference on Artificial Intelligence IJCAI 01*, pages 1195–2001. Morgan Kaufman, Los Altos, CA, 2001.

C. M. Jonker, S. Meijer, D. Tykhonov, and T. Verwaart. Multi-agent model of trust in a human game. In P. Mathieu, B. Beaufils, and O. Brandouy, editors, *Artificial Economics*, LNEMS 564, pages 91–102. Springer, Berlin, 2006.

I. Kónya. Modeling cultural barriers in international trade. *Review of International Economics*, 14: 494–507, 2006.

S. Meijer, G. J. Hofstede, G. Beers, and S. W. F. Omta. Trust and tracing game: learning about transactions and embeddedness in a trade network. *Production Planning & Control*, 17:569–583, 2006.

M. Minkov. *What makes us different and similar*. Klasika I stil, Sofia, 2007.

D. S. Wilson. *Evolution for Everyone*. Delacorte Press, New York, 2007.

Chapter 10
Agent-Based Experimental Economics in Signaling Games

Adolfo López-Paredes, Marta Posada, Cesáreo Hernández, and Javier Pajares

Abstract In this paper we built up an agent-based model inspired by human-subject behaviour in a signaling game experiment. The behavioural patterns of players observed in the experiment are classified in terms of attitudes, emotions and frugal and fast heuristics. We build up an agent-based model where artificial agents are endowed with these cognitive patterns by means of an endorsement scheme. We show the relationship between cognition and Multi-Agent interaction. We validate the model since it reproduces well the behavioural patterns observed in the human-subjects experiment.

10.1 Three Approaches to Study Signaling Games

A signaling game is a two-player game of incomplete information in which one player is informed and the other is not. The informed player sends a signal to the uninformed player who takes an action.

There is an enormous amount of literature that uses signaling models in economic applications. The earliest work was Spence's (1973) model of educational signaling. Riley's (2001) survey contains an extended discussion of some of the most important applications (advertising, limit pricing, bargaining, finance and reputation).

There are three approaches to study this problem: Game Theory, Experimental Economics and Agent-based Computational Economics (ACE). In Fig. 10.1 we have shown the way each approach analyzes it.

Game theory uses mathematics to design rules that capture the agents' behaviour providing a formal language to study how players should send and interpret signals. It asks *"What actions would be chosen by players facing these rules?"* If the results are similar to the behavior of real-life decision-makers, the model helps us

A. López-Paredes, M. Posada, C. Hernández, and J. Pajares
Valladolid INSISOC, Valladolid University, Paseo del Cauce s/n 47011 Valladolid, Spain, e-mail: adolfo@insisoc.org, [posada, cesareo, pajares]@eis.uva.es

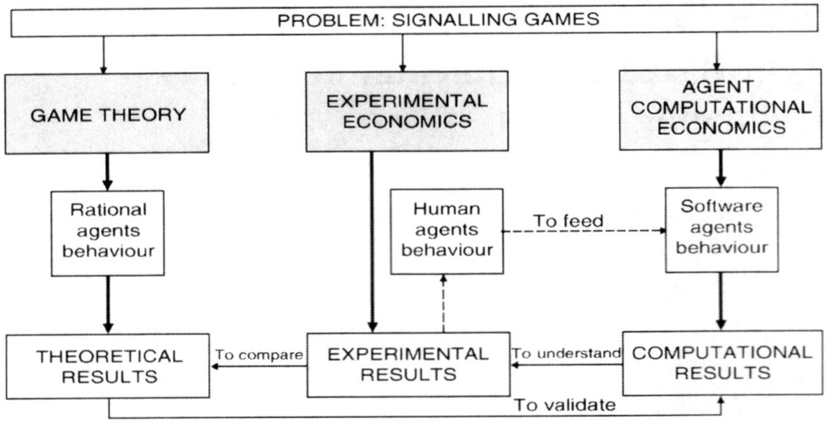

Fig. 10.1 Approaches to study signaling games

"understand" that behavior. Kreps and Sobel (1994) and Sobel (2007) review the basic theory of signaling and discusses some applications.

Experimental Economics observes human behaviour in a controlled ceteris paribus environment. It has provided evidence that human-subjects lack the strategic sophistication assumed in equilibrium models. Further, agents may be motivated by more than their material well being. Existing experimental researches (Banks et al., 1994; Brandts and Holt, 1992) provide broad support for many of the qualitative prediction results of the theory, but also suggest ways in which the theory may be inadequate.

ACE goes beyond Experimental Economics controlling the individual agents' behaviour too (López et al., 2002). López et al. (2002) claimed that both Experimental Economics and ACE methodologies complement and benefit each other. The interest in combining both methodologies is increasing in the last years. Duffy (2006) surveyed how ACE methodology has been used to understand experimental findings with human subjects and to validate ACE assumptions and simulations findings.

But also, human-subject experiments can provide a ready-made source of empirical regularities that can be used to model the individual artificial agents' behaviour. This relationship between Experimental Economics and ACE methodologies (shown in Fig. 10.1 using a dashed line) is the scope of the paper.

In this paper we use observed individual behavior in human-subject experiments to model the artificial agents's behaviour. We ask the human-subjects in a signaling game experiment about their behavioral elements (attitudes, emotions and heuristics). Then we build up an agent-based model inspired by these human behavioral elements (attitudes, emotions and heuristics). We validate it comparing the aggregate patterns observed from human-subject experiments with our simulation findings.

In this way we join Simon's (2001) critique of those avoiding micro evidence to create models: *Armchair speculation about expectations, rational or other, is not a satisfactory substitute for factual knowledge as to how human beings go about anticipating the future, what factors they take into account, and how these factors, rather than others, come within the range of their attention.*

The rest of this paper is organized as follows. In Sect. 10.2 we describe the human agents' behaviour observed from the human-subject experiment. In Sect. 10.3 we describe the artificial agent's behaviour inspired by human-subjects' behaviour. In Sects. 10.4 and 10.5 we report the parameters of the simulation and some selected results. Finally, in Sect. 10.6, the conclusions.

10.2 Human-Subject Behaviour in a Signaling Game Experiment

The two stage game goes as follow. Player 1 receives an external information about the quality level of the item to be contracted. It can be high (H) with a probability equal to 2/3 or low (L) with a probability equal to 1/3 . Player 1 sends a signal (A or B) to player 2 who takes an action (C or D). The payoff matrix used is shown in Fig. 10.2.

The result of the human-subject experiment indicates that the most repeated decisions are B-D (36%) and A-C (32%). The payoff matrix and the probability of each level are essential to calculate the equilibrium solution. But we are more interested in the behavioral elements than in the equilibrium solutions.

It was (initially) quite surprising that nobody calculated their own and the opponent's expected rewards using probabilities. People have difficulties in reasoning intuitively about probabilities.

Some of the heuristics (independent of motivations) used by human-subjects in experiments were the following:

Player 1 can initially take a selfish attitude, playing the following strategy: *I choose A when the quality level is low* and *I choose B when the quality level is high*. It would return to him the maximum possible reward. But sometimes, such an attitude will induce player 2 to play *"strategically" C and/or D.*

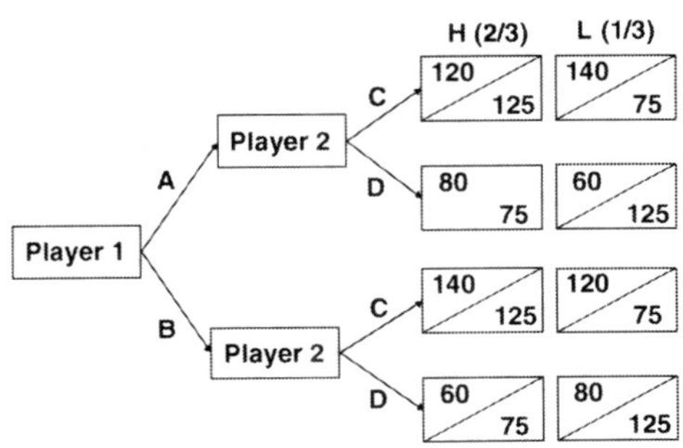

Fig. 10.2 Payoff matrix

For instance, if player 2 "accepts" player 1's strategy, he will play the best option for the global output (altruist): *I always choose C*. It is consistent because the total surplus is the maximum.

But player 2 can also play a strategy "to punish" and modify the player 1 selfish choice: *If player 1 chooses A, I choose D* and *If player 1 chooses B, I choose C*.

Some players 2 made a particular (maybe perverse) interpretation of the payoffs. They think that they can beat player 1 in each round if they *always play D*. Even more they will have the extra satisfaction to go over player 1's payoff, who has more information and is initially favored by the payoff matrix.

In these examples we can observe different players' attitudes: looking for the common benefit or for their own benefit, even more, looking for the lowest rewards of his opponent. We can observe that human-subjects change their choice reacting to the opponent's attitude and this change depends on his own attitude. In the next section we describe how we have modelled this human-subject behaviour.

10.3 Modelling Artificial Agents' Behaviour in Signalling Games

In this section we describe the agent-based model[1] (Fig. 10.3) where the behavioral elements (attitudes, heuristics and emotions) observed from the human-subject experiments are considered.

Player 1 receives external information about the quality level (H or L) and he sends a signal (A or B) to player 2 who takes an action (C or D). Each player makes his choice using a heuristic strategy which is chosen considering the rewards, his own attitude, the beliefs about the opponent's attitude and his own emotional state. We adopt the endorsements proposed by Cohen (1985) to model this behaviour.

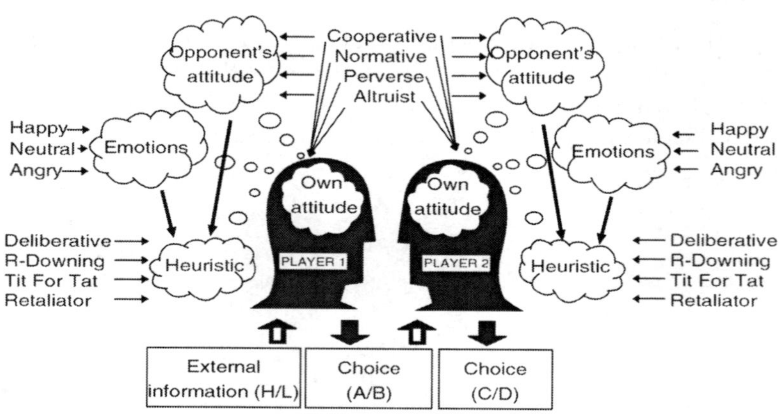

Fig. 10.3 Mental processes of artificial players

[1] The model was initially programmed in SDML, a Strictly Declarative Model Language. An applet in Netlogo will be available in www.insisoc.org/signaling-game.

See Pajares et al. (2004) for a detailed description applied to industrial dynamics problems.

An endorsement is a data structure that summarizes the reasons to believe or not in the propositions the endorsement is related to. Thus, endorsements give us a measure of the degree of confidence we have about the certainty of a proposition.

Not all the reasons to believe in something have the same (e) weight in the final certainty about the proposition. As a consequence, a hierarchy of classes has to be defined, so that, endorsements in the same class have the same degree of certainty. Furthermore, a scheme of weights must be established, in order to assess the number of the lower class endorsements that are equivalent to one endorsement of a higher class. It is called (b) base level. We have chosen a base value of 2 which indicates that an endorsement of a higher class is equivalent to two endorsements of a lower class.

Finally, a measure (E) of the belief is constructed, based on the total endorsement, which is computed considering all the weights of all the classes associated with the propositions that are true. In the case of negative weight, we must subtract the powered quantity instead of adding it to the total endorsement value. It is calculated using the following equation:

$$E = \sum b^e - \sum b^{|e|}.$$

(a) *Attitudes*. We labelled and classified the social attitudes of the human-subjects in four groups following Cesta et al. (1996). Agents can be: altruist, cooperative, normative and perverse.

- An *altruistic* player plays the best options for both players.
- A *cooperative* player is ready to partially lower his rewards in order to obtain a better sharing of the total payoff and to achieve a stable cooperation.
- A *normative (egoistic)* player only looks for agreements when these can bring any extra benefits.
- A *perverse* player enjoys damaging the opponents. He achieves less satisfaction from economic earnings than from harming his opponent.

In our model each player has the same attitude all along the simulations and does not know the opponent's attitude.

Each agent forms a belief about the opponent's attitude using the following endorsement scheme:

- The relationship between the quality level of the item and the message sent by player 1.
- The relationship between the message sent by player 1 and the action taken by player 2.
- The relationship between players' decision and the actual own rewards.

(b) *Emotions*. We adopt emotions as mechanisms which allow agents to make adaptive choices. The emotional state (satisfied, angry, neutral) is heavily endorsed to maintain or change the agents heuristic. The facts which can change the emotional state of a player are: one's own attitude and the rewards achieved related to the expected value.

The expected value depends on his own attitude (altruist, cooperative, normative or perverse). In each round each agent compares the achieved reward with the expected value. The emotions endorsement scheme is:

- The actual reward is lower than expected rewards (weight $= -1$).
- The actual reward is higher than expected rewards (weight $= 1$).

The influence of the own attitude is the following:

- An altruistic player does not alter his own emotional state.
- The cooperative player's emotional state will be affected by "opportunistic" behaviour of his opponent.
- The normative player's emotional state will be affected by the relative distribution of earnings.
- The perverse player's emotional state changes when the opponent loses less or more than he does.

(c) *Heuristic*. The human-subject behaviour confirms that individuals make decisions on the basis of heuristics, even in the face of statistical evidence suggesting alternative decision pathways. The reason is that people have difficulties in reasoning intuitively about probabilities.

Social simulation has widely benefited of the seminal work by Gigerenzer and Goldstein (1996). They demonstrated that cognitive mechanisms founded on a simple psychological mechanism (one-reason decision making) were capable of successful performance in the real world. We followed their proposal to discover the fast and frugal heuristics that participants in the experimental sessions applied in the tournaments of the game, and their motivations to choose and change from one to another heuristic.

We have considered four heuristics:

- *Deliberative*. Taking the same choice of the previous round (waiting and seeing).
- *Reactive Downing*. Taking the same choice of the previous round but this round is the last opponent's opportunity.
- *Reactive Tit-For-Tat*. Taking the same attitude that my opponent.
- *Retaliator*. Punishing the opponent.

Each agent bases the heuristic choice on his own attitude, his emotional state and the beliefs about the opponent's attitude.

The influence of own attitude is the following:

- The altruistic and the perverse players never try to change the attitude of his opponent.
- The cooperative and the normative players try to modify the other player's behaviour inducing him to a coordinated higher payoff.

10.4 Parameters and Scenarios of the Simulation

We set up the simulation sessions, to analyze two dimensions of the problem:

(a) *The effect of different attitudes on both sides of the game.* We consider 16 scenarios as result of the product of the four players' attitudes: altruist, cooperative, normative, and perverse.
(b) *The importance of the sequence of the quality level of the item received by player 1.* To control for this feature we have used four fixed quality sequences: E1, E2, E3, E4. So we have considered 64 experimental variations.

10.5 Some Simulations Results

We validate our model by comparing the simulation findings with the human-subject experimental results. Although our discussion is about the pay-off matrix of Fig. 10.2, our results can be extended considering other payoffs.

(a) *The effect of different attitudes on both sides of the game*
Two aggressive players fell nearly irremediably into a "fight" that either led to undefined situations BD, or to random choices from all the four alternatives AC/AD/BC/BD.
In all the sessions where player 2 has an altruist attitude, the players' choices end up being BC.
In all the session where player 1 has an altruist behaviour, the players' choices end up being AC/AD.
Some bargaining power for player 1 emerges playing A when his quality level is L and playing B when quality level is H, and player 2 voluntary adapts its choice to player 1. This only happens if player 2 behaves altruistically, while player 1 is either normative or perverse.
Some bargaining power for the player 2 emerges playing D when the quality level is L and playing C when the quality level is H, and P1 voluntary adapts her choice to that of player 2. This happens when player 1 behaves altruistically while P2 is either normative or perverse.
If player 2 becomes angry (being cooperative, normative or perverse), he endorses the heuristic "punishment," that means repetition of the D decision, independent of player 1 choice or of the quality level.
(b) *The importance of the sequence of the quality level of the item received by player 1.*
If the quality sequence is too variable, and either of the player 1 or the player 2 reaches a negative emotional endorsement, the game evolves with player 2 repeating taking D whilst player 1 also repeats the same decision. Both players lose interest in obtaining a good result and they focus on beating their opponent.
In Fig. 10.4 we show some results of the simulations with normative-normative couples of artificial agents, in different sequences E1/E2/E4:

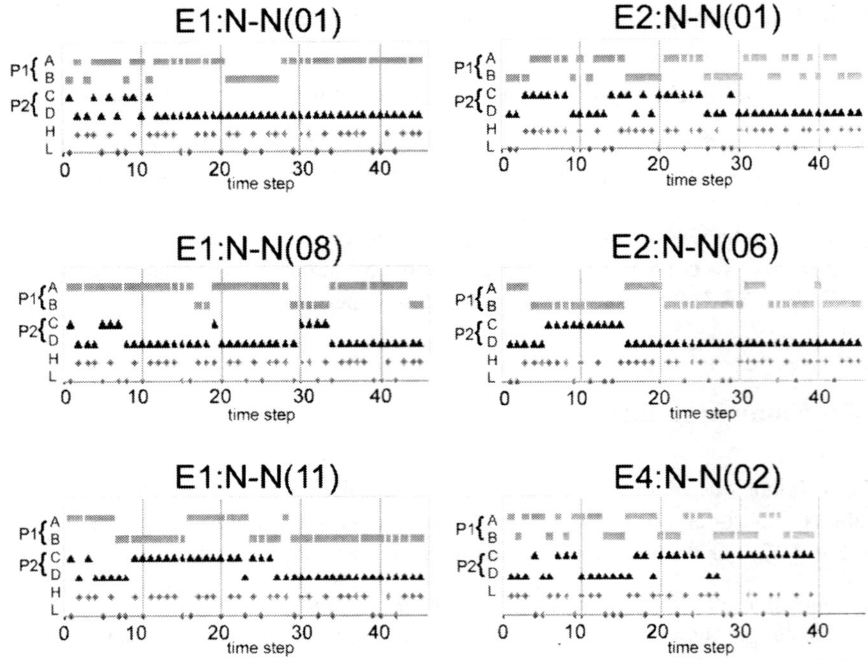

Fig. 10.4 Different sequence of the quality level in a game with couple normative-normative agents

E1-NN (01): There is a permanent AD sequence (tacit agreement) due to player 2 starting to punish player 1 from the 12th round.
E1-NN (08): The prevalent couple of choices is AD, but it is not permanent at 40th round.
E1-NN (11): Both players fix their strategies to BD from the 29th round. Both behave punishing each other.
E2-NN (01): There is no permanent couple of choices, although P2 decides to punish player 1 from 29th round. Player 1 alternates A and B.
E2-NN (06): Similar to E1-NN (11), but from the 16th round.
E4-NN (02): The sequence H/L is quite regular. Both players tacitly agree to play AC/BC from the 20th round.

10.6 Conclusions

We have looked for the behavioral elements that are consistent with the emergence of bargaining power and a cooperative attitude either of player 1 or player 2. We have built up an agent-based model where artificial agents are endowed with the cognitive patterns observed in the human-subjects experiment, in terms attitudes, emotions and frugal and fast heuristics. To this end we have used an endorsement scheme to link cognition and multi-agent interaction.

We have demonstrated that artificial agents with social and cognitive mechanisms can behave as strategically as human-subjects do in experimental sessions. Artificial agents should be endowed with the following behavioral elements attitudes, strategies and emotions.

Both emotions and heuristics can be efficiently represented using endorsements. The use of endorsements allow us to enrich and extend the artificial agent model to include cognition and emotions. Human-subject experiments are fundamental to determine the endorsements and the heuristics that finally conform the cognitive capacities of artificial agents.

The emergence of bargaining power either of the player 1 or player 2, and alternatively, the "cooperation" among artificial agents are issues that mainly depend on the relative behavioral attitudes on both sides and on the regularities (patterns) in the quality of the exchanged good, a path dependence property.

Acknowledgements This work has received financial support from the Spanish MEC, n° 2005-05676. We wish to thank two anonymous referees for their helpful comments.

References

J. Banks, C. Camerer, and D. Porter. An experimental analysis of nash refinements in signaling games. *Games and Economic Behavior*, 6(1):1–31, 1994.

J. Brandts and C. Holt. An experimental test of equilibrium dominance in signaling games. *American Economic Review*, 82(5):1350–1365, 1992.

A. Cesta, M. Miceli, and P. Rizzo. Effects of different interaction attitudes on a multiagent system performance. In *Lecture Notes in Computer Science*, volume 1038, pages 128–138. Springer, 1996.

P. Cohen. *Heuristic Reasoning About Uncertainty: An Artificial Intelligence Approach*. Pitman, Boston, 1985.

J. Duffy. Agent-based models and human subject experiments. In Tesfatsion and Judge, editors, *Handbook of Computational Economics 2*. 2006.

G. Gigerenzer and D. Goldstein. Reasoning the fast and frugal way: models of bounded rationality. *Psycholical Review*, 103:650–669, 1996.

D. Kreps and J. Sobel. Signalling. In Aumann and Hart, editors, *Handbook of game theory: with economics applications*. 1994.

A. López, C. Hernández, and J. Pajares. Towards a new experimental socio-economics. complex behaviour in bargaining. *Journal of Socio-Economics*, 31:423–429, 2002.

J. Pajares, C. Hernández, and A. López-Paredes. Modelling learning and r&d in innovative environments: a cognitive multi-agent approach. *Journal of Artificial Societies and Social Simulation*, 7(2), 2004.

J. Riley. Silver signals: Twenty-five years of screening and signaling. *Journal of Economic Literature*, 39(2):432–478, 2001.

H. Simon. *Models of Bounded Rationality*. MIT Press, New York, 2001.

J. Sobel. Signaling games. Technical report, UCSD, 2007. An encyclopedia article.

M. Spence. Job market signalling. *Quarterly Journal of Economics*, 87(3):355–374, 1973.

Part IV
Methodological Issues

Chapter 11
Why do we need Ontology for Agent-Based Models?

Pierre Livet, Denis Phan, and Lena Sanders

Abstract The aim of this paper is to stress some ontological and methodological issues for Agent-Based Model (ABM) building, exploration, and evaluation in the Social and Human Sciences. Two particular domain of interest are to compare ABM and simulations (Model To Model) within a given academic field or across different disciplines and to use ontology for to discuss about the epistemic and methodological consequences of modeling choices. The paper starts with some definitions of ontology in philosophy and computer sciences. The implicit and different ontology which underlies the approach of a same object of interest are discussed in the case of spatial economists and geographers. Finally, using the case of Shelling's model, we discuss the concept of "ontological test," and raise the question of the ontological compatibility between the "model world" and the "real world."

11.1 Introduction

Every Agent-based Computational Economic (ACE, see Tesfatsion and Judd, 2006) model or more generally artificial society model (Gilbert and Conte, 1995) has a specific "ontology" – to be defined later, either implicit, or sometimes explicit. The aim of this paper is to stress some ontological and methodological issues for Agent-Based Model (ABM) building, exploration, and evaluation in the Social and Human

P. Livet
CEPERC, UMR 6059, CNRS & Université de Provence, France, e-mail: livet@up.univ-mrs.fr

D. Phan
GEMAS UMR 8598, CNRS & University Paris IV - Sorbonne, France, corresponding author, e-mail: dphan@msh-paris.fr

L. Sanders
Géographie-cités, UMR 8504, CNRS & Université Paris 1 & Université Paris 7, France, e-mail: lena.sanders@parisgeo.cnrs.fr

Sciences (Phan and Amblard, 2007). A first goal of such an ontology is to help the model engineering and software design process. A second field of interest is to provide a corpus of formal descriptions of models in a language that is suited to the rigorous presentation of the assumptions, clarifying explanation, evaluation, and experimental methodology, in particular because it is necessary to reproduce the experiments (without going deeply into computational details). A third domain of interest is to compare ABM and simulations (Model To Model – Hales et al. (2003); Amblard et al. (2007)) within a given academic field or across different disciplines as well. A fourth domain concerns the ability that a given ontology gives us to discuss explicitly about the implications of the model's ontological commitment, in other words the epistemic and methodological consequences of modeling choices (Livet, 2007).

The present paper focuses on these two latter aspects. Indeed, different ontologies, linked with different models, could be in competition for the explanation of a given empirical problem. In Social Sciences, in many cases, we have insufficient ways to evaluate the relative adequacy of some competing models (Amblard et al., 2007); the approach consisting in showing how the ontological bases of different models could open new possibilities for evaluation and comparison. The examples are taken from the fields of economy (Tesfatsion and Judd, 2006) and geography (Sanders, 2007a,b). The point is then to explore what implicit ontology underlies the study and modeling of shared objects of research, to discuss the relevance of a more homogeneous semantic framework when different modeling syntaxes are used although the basic entities and relations are possibly the same, or on the contrary the necessity of distinguishing different entities and applying them to different operations. One aim is to improve dialog and exchanges between experts of a domain (hereafter called "*thematician*"), modelers and computer scientists as between researchers of different disciplines as well.

First we go briefly through some definitions of ontology in the fields of philosophy and computer sciences where ontological questioning is commonly developed, in order to ground the conceptual basis of the proposed approach (Sect. 11.2). In order to show the interest of making explicit the ontological status of the entities used in a modeling, we discuss then the implicit and different ontology which underlies the approach of a same object of interest, the spatial configuration of people and economic activities, by spatial economists and geographers (Sect. 11.3). Using the case of Shelling's model, we discuss then the concept of "ontological test," and raise the question of the ontological compatibility between the "model world" and the "real world," between what we call the thematic, the modeling and the computer frame of the simulation (Sect. 11.4).

11.2 From Ontology in Philosophy and Computer Science to Ontological Design for ABM

Ontology has been one of the leading domains of philosophy for a long time. For the contemporary philosopher Barry Smith (2003), ontology is "*the science of what is, of the kinds and structures of objects, properties, events, processes and relations in every area of reality*"; in a broader sense it refers to the study "*of what might exist.*" Then, defining an ontology consists in analyzing a domain, identifying the pertinent entities (objects, qualities, relations, processes), and the operations on these entities. Ontology puts constraints on the concepts that we are entitled to use in a domain (for example, concepts implying continuity cannot be applied to discrete unities). More recently, the term ontology has been imported in the fields related to computer science, such as software design and model engineering, artificial intelligence and knowledge management (semantic web, information architectures, data organization...). An ontology is then a specification of a conceptualization of a given domain (Gruber, 1993) and it deals, roughly speaking, with the formalization of the objects of knowledge in that domain: abstract types of entities or objects are defined, together with their relations. For the computer scientist and ABM designer J.P. Muller, defining a particular ontology for ABM is a very similar process even if the goal is not similar to those of philosophers. "*It consists in identifying and categorizing the pertinent objects and the relations between them for a given domain.*" *The model designer oriented ontology is then "a set of concepts / classes / categories / types, structured through taxonomic and semantic relations, concerning structures of individuals / objects / entities.*" Hereafter, we denote such model-design oriented ontology for ABM as the "*system's ontology*," where "system" is the ABM. There are two small differences with the ontology of philosophers: for philosophers, objects are not necessarily the basic entities (substances or processes are) and concepts are not ontological entities, but our ways to apprehend entities. The ontological entities that are necessary as a basis for concepts are qualities and properties.

According to these definitions, a first step is to create an ontology of the entities referred to and the concepts that we are entitled to apply to them in the specific academic domain in which the ABM is to be built. The point of creating this ontology for ABM is not only to capture the current knowledge in a domain, but also to facilitate model building. Indeed, because the kind of ontology that we deal with concerns the art of modeling a domain of the empirical reality, we need to specify what is a model in the Social Sciences on the one hand, and to discuss in what sense an ontology developed for model design has a possible relation with ontological, methodological and epistemic aspects of the corresponding academic domain, on the other.

According to the "semantic" epistemological framework (Suppes, Van Fraassen), a theory is related to experimental data (themselves related to reality) through models that specify the parameters of the theory, and apply it to a particular domain: theory is particularized in models that are confronted with experimental data extracting information from a domain of reality (the "object domain"). At each of these steps, we have ontological implications, and each time, ontology is more general, which allows different theories (models, etc.) to have the same ontology. A theory (more general and more abstract than models) implies an ontology, but it has a richer and more specific content. For example, physical laws may imply that we count relations as ontological entities, but laws specify these relations. Ontology just tells us what are the ontological types of the needed components, not what is their specific organization. For example, ontology can assume that there is a functional relation

$R(Sa,Sb)$ between the size S of A and the size of B, but do not tell us that S_A is $f(S_B)$ where f is a specific non linear function. Even if you would accept "structures" as entities, you do not need to tell what specific structures there are, which is the task of the theory. Different theories (and a fortiori different models) can have the same ontological furniture (to use Russells metaphor). This makes ontology useful for comparing theories and models.

The question is that of ontological compatibility between the thematic field and the ABM framework. *Ontology is used here as a test, in analogy with the problem of translation between two languages*. When two sentences in two different languages are about the same objects and their situations, these objects and their situations are our benchmark for assessing the reliability of the sentences that express the same situation in the two languages. For example, if we model social facts as the subsumption of situations under norms, and do not take cognitive agents but only social roles as members of the situations, our ontology is incompatible with a change of the situation triggered by the fact that some individuals have misunderstood the norm. In order for the two descriptions to be compatible, our ontology has to take at least cognitive processes as members of the situation.

An ontological design implies some formal specifications. First a choice must be made for the manner in which entities are to be described, then which relations need to be represented and finally how they should be conceptually represented. Roughly, that is the equivalent of a formal dictionary of this domain of interest. It will thus act as a *meta-model* for organizing and conceptualizing the knowledge in a domain by means of a formal language which is understandable not only for the thematicians and the modelers) but also for the computer scientists. As we can build different virtual worlds, we need a definition of their basic ontology in order to compare them, and to confront them to other descriptions of the social empirical reality.

11.3 From Individuals to Spatial Entities: What Entities Make Sense from the Ontological Standpoint?

The starting point to design a domain-related ABM ontology (a system's ontology) is to specify the relevant entities within the domain and the concepts that can be applied to them in a relevant way. Depending on the question under study, the starting point will differ. For example, a basic ontology in economy is based on the existence of economic agents and transactions (exchanges) between them. Taking the stock market as an example, for handling the associate mechanisms, space is not necessary and time is generally highly abstract (except in particular cases, as for instance where micro structures are explicitly taken into account, see among others (Daniel, 2006)). On the other hand, a basic ontology in geography is based on the existence of localized entities and spatial relations. Consequently, the operations defined on entities refer to their localizations. Scales may have also interesting effects: entities of different scales may not be considered as the same object (Openshaw, 1996, 1997). The models developed in each discipline are strongly influenced by these underlying principles, and even when the objects and the questions are similar, the way of treating them will differ.

For a social scientist, a first hunch on ABM suggests a strong methodological individualist background, since "agent" looks like "individual" (human). But this apparent similarity is not so relevant that it seems and could be fallacious. First, in the computer science, a "software agent" is nothing but a software technology, e.g. a specific way to design software in object-oriented paradigm, inherited from the generic class "object" but with more specified properties (see Ferber, 1999, 2007). Such software can be used for the implementation of numerous software objects, then "agentified," that are non-relevant from the model point of view (like graphic interfaces). Second, in a multi-agent framework, the agents can be the avatar of non-human active entities (like ants) or collective entities (like firms or cities) as well. In many cases, avatars of humans (individuals) do not make sense as "agents," because humans are embedded into largest entities (i.e. cities) or are only support infra-individual entities that are the subject of interest (such as functions, role, routines, genes...). Thus, the design of an artificial society does not imply that the relevant software "agents" are designed to represent individuals.

Accordingly, a non-individualist ontology is possible for multi-agent systems, even if the individualist ontological design prevails in practice in some academic fields (i.e. "individual-based" in ecology (see e.g. Grimm, 1999), or "agent based" in economy (see e.g. Tesfatsion and Judd, 2006)). This could be for instance the case in geography (Sanders, 2007a,b). Finally, in some other agents architectures, the object of the analysis (and the subject of interest as well) are not the individuals themselves. Individuals are only the support of other entities of interest, as in the models of (self-organized) emergence of shared language structures, reviewed by Hutchins and Hazlehurst (2002) or in epidemiology. Such an ontology is not reducible to the orthodox view of methodological individualism, since the entities of interest are infra individuals (Sperber, 1997) and knowledge is not a predetermined content inside the head of individuals, but the product of an autonomous process, distributed within groups (Hutchins, 1995). Depending on the hypotheses on where are the driving forces in the dynamics of the phenomenon under study, the modeler can then choose to represent with a software "agent" an infra-individual entity, an individual, or a supra-individual entity.

Geographical entities illustrate the latter case. Depending on the academic background, such entities intervene in models with a different status, what sometimes leads to contradictions that a reflection on underlying ontology could help to solve. The place of space and scales in new economic geography on the one hand, and in geography on the other, are fruitful examples. Both fields of research are concerned with the spatial organization of economy and people. But scales and space raise ontological questions. Is a spatial grid a constituent of the basic ontology, or is it a purely conceptual way of putting entities in different classes? Is space absolute and given before the entities, or is it dependent on them? Are different scales different ontological levels, or just different epistemic accesses? If physical phenomena can be different on different scales, are scales ontological boundaries? In the same way, are administrative regions relevant ontological entities?

In the new economic geography approach (Krugman, 1991; Fujita and Thisse, 2002), cities for example are interpreted as the result of the interaction between agglomeration and dispersion forces. In the theoretical models developed in this framework, spatial entities are considered as abstract points localized in a homogeneous environment. The model concentrates on the process of concentration and initial conditions will determine whether and where concentration will take place. In the theoretical model, these initial conditions are introduced with a random component. The place where it happens is not of prime importance, since the focus of analysis is on the cumulative processes of agglomeration which lead to lock in a certain pattern of evolution in a certain locality. The spirit is to model a stylized fact, as detached as possible from the context, and deepening the theoretical analysis of the mathematical model as much as possible. Some of these models are then confronted to empirical data in order to evaluate their credibility. Most of such applications use existing data observed at the level of administrative entities. The question is then to what degree these administrative entities correspond, in their sense, in their being, to the spatial entities considered in the model. Ontologically, the operations related to a spatial grid (localization, mainly) may be different from the operations related to an administrative qualification (department or municipality, for example). Gap is frequent and this question should not be underestimated as it directly influences the validation. It is here that the question of scale is central, and in particular the question of what is called the Modifiable Areal Unit Problem (MAUP).

The MAUP designs the fact that the results of a statistical treatment, of a cartographic representation, of a modeling, may be different if the units of observation correspond to one scale or another, to one zoning (or aggregation) or another. For example, the correlation between vote and income is different if these variables are observed at the level of the blocks or the quarters of a city, as well as at the level of the cities themselves. It can easily vary from significantly positive to not significant or even to significantly negative. This question has given rise to a huge amount of literature since the 1970s (Openshaw and Taylor, 1979). To simplify, it is possible to summarize the responses of researchers to this question through three categories:

1. To drop all kinds of aggregates and to only work at the level of individuals (Benenson and Torrens, 2004; Boman and Holm, 2004, etc);
2. To develop technics and methods in order to face the problem; statisticians have developed tools which integrate the MAUP, through weighting methods for example (spatially weighted regression, Fotheringham et al., 2000, etc);
3. Not to interprete this feature as a "problem" but to see it, on the contrary, as giving fuller description and understanding of multi-scalar phenomena (urban segregation for example is operating at several scales). Space and scale are then seen as playing a driving role in a societys dynamics and an important aim is then to identify whether similar or different mechanisms are at work at the different spatial scales. A discussion on the ontological status of the geographical entities associated to the concerned different scales is a necessary step for getting better understanding of these phenomena.

Geography has often an approach to modeling which is less mathematical and the analytical properties of the formal model are often poorly explored. The aim is more often to model an observed phenomenon than a stylized fact (see Sanders, 2007a), and especially to take explicitly the context into account. This context represents the environment in which a phenomenon appears and a systemic approach is often adopted to identify how sufficient and necessary conditions have combined in order to produce observed dynamics.

Both economical geography and NEG are concerned with the spatial organization of economy and people, but the models developed are of different nature in terms of formalization as well as in terms of privileged mechanisms of explanation. In fact the questioning itself is not exactly the same. As Thisse (Espace Géographique, 2007) stresses it, the relevant question in the new economic geography is "Why are there cities?" when in Geography the question is rather "why is there a city here and not there?." In that sense, the approaches seem to have complementary rather than contradictory properties.

It is even possible to show that the two ontologies which underlie respectively economical geography and NEG can be combined: agents represented by a list of preferences and a list of possible actions can be localized, can have preferences for spatial locations and spatial relations with their neighbors; spatial operations are among their possible actions. On the other hand, space is more a relation than an operation for economists, or a factor of cost in exchanges and is both relation and operation for geographers. But costs can be easily appended to the translations. A common ontology is possible but most often it is not explicitly discussed, even in comparative works.

Not only different disciplines like geography and economics may be compatible via ontology, but there can be interrelations – in all directions – between the social science realm (SS domain paradigm), the model design realm (ABM paradigm) and the software design realm (Multi-Agent System paradigm). This relationship could be sometimes non-neutral, may be dialectic. On the one hand, the choice of a particular design at the software level, or at the model level could be non-neutral for the related academic realm, embedded in some implicit or explicit commitments in all the discussed perspectives: ontological, methodological and epistemological. On the other hand, the associated commitments (ontological, methodological and epistemological) from the object domain in social sciences should imply specific architecture at the model level and/or at the software level as well.

11.4 Model vs. "Real" World and Ontological Test

Different academic backgrounds may lead to different ways of developing models, but also, and this is perhaps more interesting to be underlined in this discussion, to interpret the same models in different ways. Segregation in urban space is an relevant example of investigation shared by economists and geographers. Schelling's

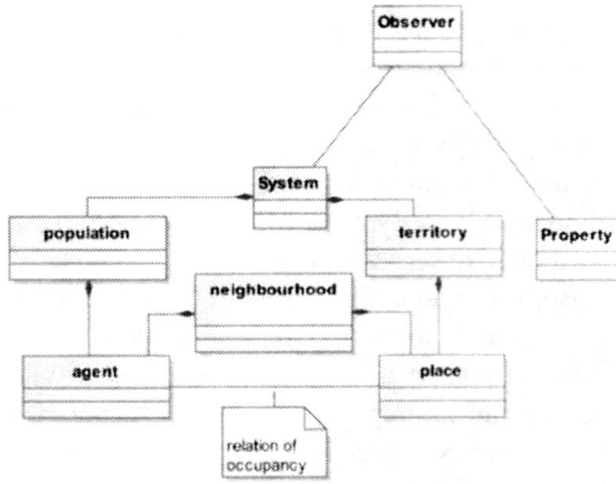

Fig. 11.1 A simple UML ontology of the basic Schellings model

model of spatial segregation (Schelling, 1978) is a source of inspiration in both fields but expectations may be different.

Schelling's aim was to explain how segregationist residential structures could spontaneously occur from local rules, without external intervention, even when people are not so very segregationist themselves. Agents are located on a checkerboard. Taking the "color" of agents as a criterion for discrimination, agents choose a location where to live, depending on their individual tolerance threshold of different colors in their neighborhood (local interaction within the "Moore" neighborhood, i.e. eight closed neighbors).

On Fig. 11.1, an ontological diagram in UML is represented for the basic Schelling checkerboard model. The system is composed of a population of agents and a territory composed of places. Agent and place are linked by a relation of occupancy: a given place is occupied by an agent or not. The neighborhood is a particular entity composed both by some places and by agents, if they are located on these places. A specific neighborhood is attached at each place (i.e. Moore neighborhood, in a particular topology - rectangular cells – of places in the territory), and characterized by the agents located at the place within the neighborhood. In the original model, the observer of global regularities (e.g. segregationist patterns) is *exterior* to the ABM system. Accordingly, this observer could be the experimentalist, thematician or modeler, but not an agent within the ABM system.

For Sugden (2002), Schelling's approach can be summarized by two claims. (1) Schelling claims that a regularity R (or stylized fact) occurs (or often occurs) in empirical phenomena: here a persistent segregation in housing. (2) This regularity *might* be explained (with parsimony) by a limited set of causal factors F, based on the simple local preferences about neighborhood (an agent moves if there are more than 66% of neighbors with different colors in his neighborhood).

In Shelling's checkerboard model, the "fully integrated structure" (each color alternates in all directions on the checkerboard) is an unstable equilibrium: a slight perturbation is sufficient to induce a chain reaction which leads to the emergence of local segregationist patterns, even in this case when individuals have a certain

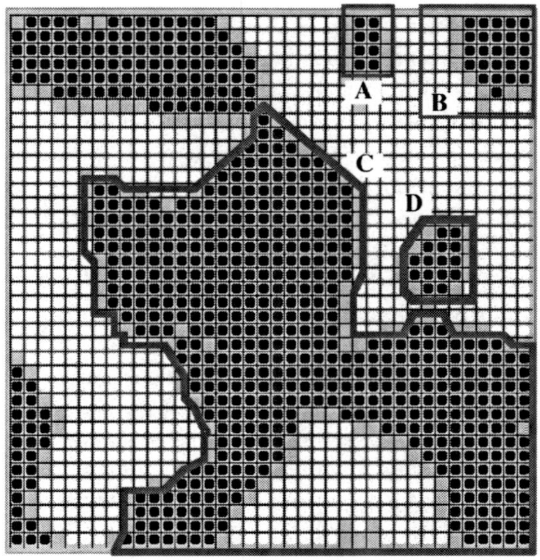

Fig. 11.2 Extended version of Schellings model. *Source*: Moduleco/MadKit

tolerance for being in minority (Fig. 11.2). The conclusion is that segregation *might* be the result of (even weak) individual preferences for living in a local neighborhood where individuals of another category than themselves do not exceed a certain threshold. For Sugden (2002), Schelling "is declaring his confidence that this approach is likely to work as an explanation even if he does not claim so to have explained anything so far. (..) He constructed imaginary cities which could be viewed as possible cities, alongside real cities. We are invited to make the inductive inference that similar causal processes occur in the real cities." The model is then accepted as a *plausible candidate* in the argumentative debate for explaining segregation.

The driving mechanism in this model (the response of individuals to the configuration of their neighborhood) is related to space, and therefore attractive for geographers. But this spare version is not adapted for a multi-level context, which is quite dominant in urban geography. The introduction of effects from different levels of neighborhoods leads to formalizations with no analytical solutions. On the other hand, the role of existing spatial entities and of existing boundaries between different entities, are often supposed to be central in the empirical world. Two neighboring cells, one at the periphery of a well off quarter, the other at the periphery of a poorer one, will for example have quite different potentials of change due to the different contexts in which they are located. At last, if a structure has emerged, it will of course influence in turn the lower levels dynamics. The geographer will even tend to give a meaning (or significance) to the associated entities (Fig. 11.2), to consider that objects exist at that level of observation, when, most often, that level is not ontologically relevant in mainstream economics. From a methodological point of view, when these features are introduced in a simulation model, most often it looses its formal properties. From an epistemological point of view, the question is that of the meaning of the objects that have emerged, and how to relate these objects to spatial entities existing at that level of observation. The final question is how these multi-scale considerations interfere with the basic mechanism of Schellings model.

To summarize, one could say that in one case (economics) the priority is to develop a model which is compatible with existing theory, and in the other case (geography) it is to develop a model which corresponds to a "realistic" representation of the world (here the multi-scalar "reality" of urban space). Considering the two points of view the question is the following: can the development of an ontology fill the gap between these different approaches? Or make explicit what transformations have to be done to go from one perspective to the other? In order to elaborate this question we have first to make explicit the operations that enable us to individualize territorial entities. The output of a simulation in using Schelling's principles in Fig. 11.2 is used to illustrate some questions about this individualization. In this example the ontological problem consists in defining the ontological status of boundaries relatively to entities; Figure 11.2 makes obvious some questions about this individualization.

In this example the ontological problem consists in defining the ontological status of boundaries relatively to entities. If we focus on forms of type A or D, our notion of entity is a domain surrounded by closed boundaries, and presenting topological connectedness. If we focus on forms of type B, our entities are space locations enclosed in a given framework - a territory. These space locations imply the existence of a relational spatial grid, if their selection is ontological, or they are just conventional locations, if the blue grid is conventional. If we focus on forms of type C, we individualize as entities forms connected by a boundary that is made by the contrast between two opposite patches (one white and one black). The entity will be a relational one, but the spatial relation will supervene on the complementation relation, and will depend on it.

The relation between ontology and methodology, described here in the case of ABM can take at least two general forms, depending on the point of view: pragmatic or realistic. From a pragmatic point of view, goal-oriented methodology should determine ontology which has to be adapted in an optimal way to the supposed nature of the procedures and experimental devices considered as effective. We do not need to endorse a realistic point of view, according to which ontology should determine the methodology which has to be adapted in an optimal way to the supposed nature of the entities regarded as existing in the reality. We have not to try to define a common basic ontology, but only to use as many different formal representations as we need for our pragmatic purposes.

Every individualization of a territory is a good one but for a specific purpose. But then we would have difficulties to compare the merits and disadvantages of different representations and also to define the articulation between their operations and elements. Even from a pragmatic perspective (for the purpose of comparative evaluation and combination of models) it is better to make the effort of making explicit what a common ontology could be.

But then we would have difficulties to compare the merits and disadvantages of different representations and also to define the articulation between their operations and elements. Even from a pragmatic perspective (for the purpose of comparative evaluation and combination of models) it is better to make the effort of making explicit what a common ontology could be. In other words, a pragmatically-oriented methodology, corresponding to a specific scientific project within a particular academic context presupposes, explicitly or implicitly, a related ontology with its own goal, relevance, and limits. The key question is then about the relevance of this abstract and fictitious "model world" for the explanation of related empirical phenomena in the "real world." The recognition of this problem is widely accepted among scientists, but there are many ways to answer the problem (Sugden, 2002).

In social sciences, different ontologies, linked with different models are often in competition for the explanation of a given empirical problem; in many cases, we have no way to evaluate empirically the relative adequacy of these competing models. But we have ways to show that different models have or not a compatible ontology. Even if we cannot claim that our ontology is the one of the real world, it will be a fruitful tool for comparing models. An ontological question makes sense at first within a particular conceptual framework, subordinated to a principle of internal consistency. But this ontological question can be an overlapping with a different conceptual framework (as ontology can be coarser grained than concepts): model world and "real world" can be compared. In economics for instance, "conceptual exploration" (Hausman, 1992) focuses on the internal properties of the model itself, without taking into account the question of the relationship between the "model world" and the "real world." For Sugden (2002) there is an irreducible gap between these two worlds. The model world is only a "possible reality" (a condition of possibility) that need to be viewed as "credible" one. Accordingly, ontology is an argument among others to do so, i.e. that can be used as benchmarks to compare ABM.

11.5 Conclusion

We have shown that it is useful to ask ontological questions and to try to define the ontological commonalities and differences between several descriptions, theories and ways of modeling. Indeed, when we build a model, the elaboration of the associated ontology can start from different points of view, that of the philosopher (epistemological grounds), of the specialist of the academic domain (who relates discourse to conceptual modeling), the modeler (from conceptual to formal modeling), or the computer scientist (who stands for the implementation). The "ontological test" consists in verifying that whatever the starting point, the final construction is the same. It means to verify the compatibility and consistency between the different points. That way, it will be possible to compare different descriptions and models, to better understand their relations and to integrate diverse models and levels of analysis in a common framework.

Acknowledgements The authors acknowledge the program ANR "CORPUS" of the French National Research Agency for financial support through the project "COSMAGEMS" (Corpus of Ontologies for Multi-Agent Systems in Geography, Economics, Marketing and Sociology). We also gratefully acknowledge Jean Pierre Müller for his significant intellectual contribution and to Alexandra Frénod for significant corrections on the original draft, P.L, D.P. and L.N are CNRS member.

References

F. Amblard, P. Bommel, and J. Rouchier. Assessment and validation of multi-agent models. In D. Phan and F. Amblard, editors, *Multi-Agent Modelling and Simulation in the Social and Human Sciences*, chapter 4, pages 93–114. Bardwell Press, 2007.

I. Benenson and P. M. Torrens. *Geosimulation; Automata-based modeling of urban phenomena*. Wiley, New York, 2004.

M. Boman and E. Holm. Multi-agent systems, time geography, and microsimulation. In M-O. Olsson and G. Sjöstedt, editors, *Systems Approaches and their Applications*, pages 95–118. Kluwer, Dordecht, 2004.

G. Daniel. *Asynchronous Simulations of a Limit Order*. PhD thesis, University of Manchester, UK, December 2006.

Espace Géographique. Débat: Nouvelle économie géographique et géographie: quel dialogue? *LEspace Géographique*, 36(3):193–214, 2007.

J. Ferber. Multi-agent concepts and methodologies. In D. Phan and F. Amblard, editors, *Multi-Agent Modelling and Simulation in the Social and Human Sciences*, chapter 1, pages 7–34. Bardwell Press, 2007.

J. Ferber. *Multi-agent Systems: an Introduction to Distributed Artificial Intelligence*. Addison Wesley Reading, MA, 1999.

A. S. Fotheringham, C. Brunsdon, and M. Charlton. *Quantitative Geography: Perspectives on Spatial Data Analysis*. Sage, London, 2000.

M. Fujita and J. F. Thisse. *Economic of agglomeration. Cities, Industrial Location and Regional Growth*. Cambridge University Press, Cambridge MA, 2002.

N. Gilbert and R. Conte, editors. *Artificial Societies: The Computer Simulation of Social Life*. UCL Press, London, 1995.

V. Grimm. Ten years of individual-based modelling in ecology: what we have learned and what could we learn in the future? *Ecological Modelling*, 115:129–148, 1999.

T. R. Gruber. Toward principles for the design of ontologies used for knowledge sharing. In N. Guarino Poli, editor, *International Workshop on Formal Ontology*, Padova, Italy, 1993. Published in *International Journal of Human-Computer Studies*, 43(5–6), 1995, p 907–928.

D. Hales, B. Edmonds, and Rouchier J. Model to model analysis. *Journal of Artificial Societies and Social Simulation*, 6(4), 2003.

D. M. Hausman. *The Inexact and Separate Science of Economics*. Cambridge University Press, Cambridge, MA, 1992.

E. Hutchins. *Cognition in the Wild*. MIT press, Cambridge, MA, 1995.

E. Hutchins and B. Hazlehurst. Auto-organization and emergence of shared language structure. In A. Cangelosi and D. Parisi, editors, *Simulating the Evolution of Language*, pages 279–305. Springer, London, 2002.

P. R. Krugman. Increasing returns and economic geography. *Journal of Political Economy*, 33(2): 483–499, 1991.

P. Livet. Towards an epistemology of multi-agent simulation in social science. In D. Phan and F. Amblard, editors, *Multi-Agent Modelling and Simulation in the Social and Human Sciences*, chapter 8, pages 169–194. Bardwell Press, 2007.

S. Openshaw. Using a geographical analysis machine to detect the presence of spatial clusters and the location of clusters in synthetic data. In P. Boyle and F. E. Alexander, editors, *Methods for Investigating Localised Clustering of Disease*, volume 135, pages 68–87. IARC Scientific Publication, Lyon, France, 1996.

S. Openshaw. Developing gis-relevant zone-based spatial analysis methods. In M. Batty and P. Longley, editors, *Spatial Analysis: Modelling in a GIS Environment*, pages 55–73. Wiley, New York, 1997.

S. Openshaw and J. Taylor. A million or so correlation coefficients: three experiments on the modifiable areal unit problem. In Wrigley, editor, *Statistical applications in the Spatial Sciences*, pages 127–144. Pion, London, 1979.

D. Phan and F. Amblard, editors. *Agent-based Modelling and Simulation in the Social and Human Sciences*. Bardwell Press, Oxford, 2007.

L. Sanders. Agent models in urban geography. In D. Phan and F. Amblard, editors, *Multi-Agent Modelling and Simulation in the Social and Human Sciences*, chapter 7, pages 147–168. Bardwell Press, 2007a.

L. Sanders, editor. *Models in Spatial Analysis*. Geographical Information Systems series. ISTE, London, 2007b.

T. S. Schelling. *Micromotives and Macrobehaviour*. Norton and Co, New York, 1978.

B. Smith. Ontology. In L. Floridi, editor, *Blackwell Guide to the Philosophy of Computing and Information*, pages 155–166. Blackwell, Oxford, 2003.

D. Sperber. Individualisme méthodologique et cognitivisme. In R. Boudon, F. Chazel, and A. Bouvier, editors, *Cognition et sciences sociales*, pages 123–136. Presse Universitaires de France, Paris, 1997.

R. Sugden. Credible worlds: The status of theoretical models in economics. In U. Mäki, editor, *Fact and Fiction in Economics: Models, Realism and Social Construction*, chapter 5, pages 107–136. Cambridge University Press, Cambridge MA, 2002.

L. Tesfatsion and K. L. Judd. *Handbook of Computational Economics, Vol. 2: Agent-Based Computational Economics*. Elsevier North-Holland, Amsterdam, New York, 2006.

Chapter 12
Production and Finance in EURACE

Sander van der Hoog, Christophe Deissenberg, and Herbert Dawid

Abstract EURACE is a major FP6 STREP project aiming at constructing an exhaustive agent-based model of the European economy, populated by a very large number of sophisticated, autonomous agents. The EURACE model, which has an explicit spatial structure, includes all the major markets considered in quantitative macroeconomic modelling (consumer goods, investment goods, labour, credit and finance). It offers a unique opportunity for studying, from a new perspective, the empirically observed but theoretically poorly understood link between the real and the financial sphere of a modern economy. After summarily presenting the main features of EURACE, this paper describes in more detail the newly developed *financial management module* that intermediates between the real and the financial spheres in EURACE. In a nutshell, this module defines the link between the hiring and investment behavior of the firms as a function of the revenues they obtain by selling their products, of the money they can raise on the credit and financial markets, of their dividend policy, and other major aspects of financial decision-making.

S. van der Hoog
Université de la Méditerranée II and GREQAM, Château Lafarge, Route des Milles, 13290 Les Milles, France, e-mail: svdhoog@gmail.com

C. Deissenberg
Université de la Méditerranée II and GREQAM, Château Lafarge, Route des Milles, 13290 Les Milles, France, e-mail: christophe.deissenberg@univmed.fr

H. Dawid
Bielefeld University, Dept. of Business Administration and Economics, Universitätsstrasse 25, D-33615 Bielefeld, Germany, e-mail: hdawid@wiwi.uni-bielefeld.de

12.1 Introduction

An important question in economics is to explain the link between the real and financial sphere of the economy. Empirically speaking, the macroeconomic and financial variables are correlated along the business cycle. The existing theoretical models, however, have difficulty explaining the observed correlations.

This paper documents recent work carried out to address this problem from an agent-based perspective. This work is part of a major FP6 STREP project, EURACE, that aims at developing a comprehensive agent-based model of the European economy. In this paper, we sketch the broad structure of the EURACE model and present in more detail the model of financial policy decisions by the firm that links the real and the financial spheres within the model.

This paper is organized as follows. In the second section, we succinctly present the EURACE project. The third section presents the financial management module. A final fourth section concludes.

12.2 The EURACE Project

Most of the existing agent-based models cover only a single industry, one restricted geographical area, or a unique market, and involve relatively small populations of agents. In contrast, EURACE aims at creating an integrated agent-based model of the European Union, linking all the various markets one typically encounters e.g. in a large dynamic CGE model of a multi-national economy. The thus defined artificial economy is to be populated with a very large number (possibly tens of millions) of fairly sophisticated agents.

The model has an explicit spatial structure mimicking the regional statistical units used by Eurostat. It includes various (typically, regional) artificial markets for real commodities (that is, consumption goods, investment goods and labour), and markets for financial assets (such as loans, bonds and stocks). For a general overview of the EURACE model, see Deissenberg et al. (forthcoming).

The three-years EURACE Project started in September 2006. It includes economists and computer scientists from eight research centres in Italy, France, Germany, the UK, and Turkey, as well as the 2001 Nobel laureate in economics, Joseph Stiglitz. More institutional and scientific details can be found on the project's web page: `www.eurace.org`.

By its scope and complexity, the effort is unsurpassed and needs to cover much terra incognita, among others concerning the conceptual and computational architecture of the model, its numerical implementation, its validation, and the exploitation of the simulation results. In particular, running such a large model will necessitate using massively parallel computing clusters, using pioneering software such as the *Flexible Large-scale Agent Modelling Environment* (FLAME).

12.2.1 FLAME

Developed by Simon Coakley, Mike Holcombe, and others at the University of Sheffield (see `www.flame.ac.uk` for a more complete presentation and references), FLAME provides a computational framework allowing modellers to easily create, exchange, include and couple models written in a high-level modelling language. Other important aspects include the development of parallelisation techniques, the distribution of agents over many processors, and the inclusion of testing methods to verify developed models. All these elements are vital to agent-based models in general and to EURACE in particular. The framework has been adapted to enable it to run on parallel computing platforms. It has been previously used to study the behavior of a number of biological systems – at the molecular, cellular, tissue and social levels – and has been instrumental in uncovering a number of new biological properties that have been confirmed experimentally by Coakley et al. (2006).

The approach followed in FLAME is to define each agent as a so-called *Stream X-Machine* (Laycock, 1993). That is in a nutshell: as an automaton described by a finite number of states, transitions between those states, and actions, and endowed with both an internal memory and the ability to interact by sending and receiving *messages*.

In the FLAME framework, agents act within *contexts*, that is preeminently, within markets. The agents can have different *roles* in different contexts, e.g., an agent can be a buyer on one market, a seller on another. Table 12.1 lists the main classes of agents in EURACE, the contexts in which they operate and the main messages they exchange.

Table 12.1 Main agents, contexts, roles, and messages in the model

Agent	Context	Role	Messages
Household	Consumption goods market	Buyer	units demanded
	Labour market	Worker	application, accept/reject job
	Credit market	Depositor	cash holdings
	Financial market	Investor	stock/bond orders
Firm	Investment goods market	Buyer	units demanded
	Consumption goods market	Seller	price, quality
	Labour market	Employer	vacancy, job offer
	Credit market	Borrower	loan request
	Financial market	Issuer	stock/bond orders
Investment Goods Firm	Investment goods market	Seller	price, productivity
	Labour market	Employer	vacancy, job offer
Bank	Credit market	Lender	credit conditions
Government	Financial transactions		tax payments
Central Bank	Credit market	Regulator	

12.2.2 The Real Sector

Somewhat simplifying, the real sector is composed of an investment goods sector, a consumption goods sector, and a labour market. The investment goods sector, however, is not agent-based, but modelled as a passive entity whose behavior is determined by simple rules. In a nutshell, it provides an infinite supply of investment goods at exogenously given prices. The productivity of the investment goods increases over time according to a stochastic process. The amounts paid for the investment goods are channeled back into the economy.

Together with labour, the investment goods are used in the consumption goods sector to produce consumption goods. These goods are sold to the households. The consumption goods firms (the firms thereafter) follow plausible, to the largest possible extend empirically grounded rules for investment, production, stocking, pricing, hiring and firing, dividend payment and/or debt making. Likewise, the households follow plausible rules for saving, consuming different types of products, looking for another job while being employed or trying to find a job while unemployed.

The workers are characterized by (i) a general skill level; and (ii) specific skills. The specific skills are acquired on the job to fully exploit the technological potential of the physical capital being used in the production process. The general skill level is obtained through schooling. The higher the worker's general skill level, the faster it acquires the specific skills associated with a given job.

To capture spatial effects, the economy is divided into regions. Consumption occurs locally within each of these regions. The firms send the goods they produce to diverse local outlet malls. The local prices may differ from one outlet mall to the next, i.e. from region to region. Households and firms are distributed across the regions and are allowed to migrate. Workers can apply for jobs in any region, but working outside of their own region of residence is associated with commuting costs that have to be subtracted from the wage.

12.2.3 The Real-Financial Interaction

Figure 12.1 sketches the place of the financial management in the interaction between the real and financial spheres.

The financial management module serves to reconcile the production and investment goals of the firm with its payout policy and other financial commitments. The firms receive revenues from the sales on the consumption goods market. This income is used to remunerate the workers they hire on the labour market and to buy investment goods. It is also used to pay the firm's prior financial commitments, i.e. to service the debt (interest and debt installment payments) and to pay taxes. Part of it may also be used to pay out dividends or to repurchase shares on the financial market. Alternatively, the firm can borrow on the credit market and/or issue stocks on the financial market to raise financial capital.

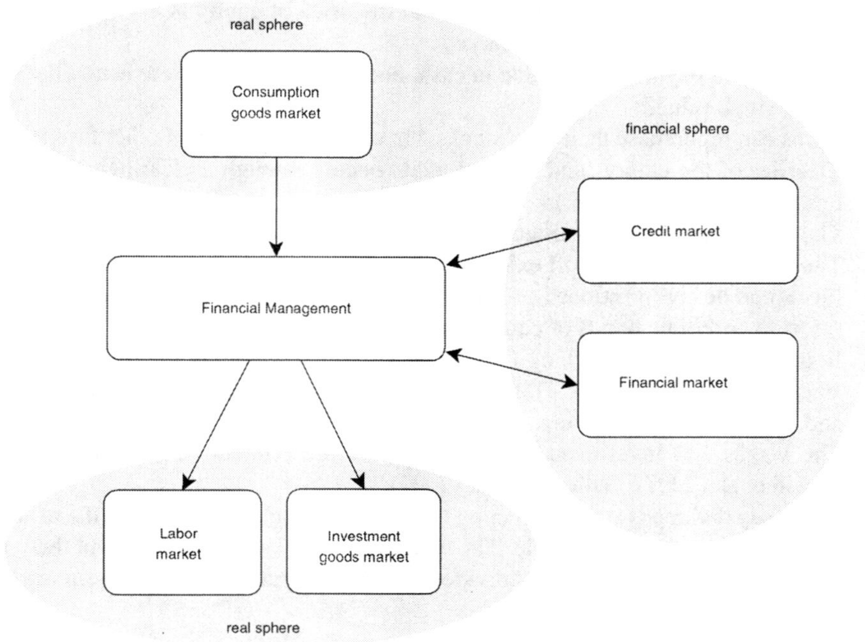

Fig. 12.1 Financial management as intermediary between the real and financial spheres

12.3 The Financial Management Module

All firms in EURACE are corporations. Therefore we use insights from corporate finance and capital management theory to model their financial decision-making. The behavioral rules are kept as close as possible to those used in reality. We define them based, among others, on the seminal contribution by Cyert and March (1963/92) on a Behavioral Theory of the Firm and the work of Myers (1984) on Pecking Order Theory, a behavioral theory of corporate finance. We also draw information from Tirole (2006, Ch. 2), Allen and Michaely (2004), and Brav et al. (2005), who report empirical evidence on payout policies.

12.3.1 General Assumptions

The following general assumptions are captured by the financial management module.

- The firm's production plan is chosen before the financial policy is decided on. Thus, in the case of newly issued equity, the new shareholders do not influence the production decision for the current period. Hence, preferred stocks and

common stocks are indistinguishable and the price of equity is the same on the primary and secondary equity markets.
- All dividend payments are made in cash, not in stocks. Thus, there is no dilution of the stock value.
- Firms can repurchase their own stocks, but cannot buy stocks of other firms.
- Clearing of the equity- and bond markets occurs through a clearinghouse or a limit-order book mechanism.
- The credit and capital markets are imperfect and subject to transaction costs. Thus, a non-trivial trade-off exists between different financing instruments.
- Firms can be credit rationed on the credit market and consequently may need to resort to corporate bonds or equity for financing.
- In accordance to the Pecking Order Theory, the firm prefers internal financing since it is the least risky. This is followed by debt financing, corporate bonds and, finally, equity financing.
- The wages and investment costs are paid before production takes place, see Dawid et al. (2007); Gallegati et al. (2007).
- Firms pay dividends at the beginning of each operating cycle, based on the results of the previous operating cycle. The total dividend payout may vary, but there is an effort to maintain a constant dividend to earnings ratio (see, e.g. the survey by Brav et al., 2005).

Over the course of the business day (one iteration) the markets open and close in the following sequence:

- Credit market
- Financial market
- Labour market
- Investment goods market
- Consumption goods market

Each firm is active on the first four markets during one day every month. This activation day varies from firm to firm and is chosen randomly at the start of the simulation. During their activation day, the active firms complete a so-called operating cycle, as described below. The selling of output however occurs over the course of the entire month that follows the activation day.

12.3.2 The Operating Cycle

During its operating cycle, the firm executes four main routines:

1. Production planning routine
2. Financial planning routine
3. Production routine
4. Accounting routine

It also checks at different points whether or not it is bankrupt, as explained in more detail later on.

Production Planning Routine

The operating cycle starts with production planning: the firm calculates the production level it would like to realize, the workforce it would like to hire, and the investments in new physical capital it would like to purchase in the absence of financial constraints. It computes the corresponding estimated costs. It also decides upon the dividends to be paid. However, it does not contractually commit to hire respectively to buy or to pay dividends since at this point of its decision-making process it is not sure that the required financing will be available.

Since, as previously assumed, the wages and investment costs are paid before the production takes place, the firm may need to obtain a bank loan or other external capital before being able to start producing.

Financial Planning Routine

After completing the production planning routine, the firm computes all financial commitments from the previous period: interest payments, debt installment payments. These, together with the production and investment costs and the dividend payout previously calculated, determinate the total liquidity needs for the current operating cycle. These needs are to be financed by internal and external means before production starts.

As previously stated, the firm's first attempt is to use internal financing since this is the least risky alternative. If the current internal financial resources are insufficient they try to complete them on the credit market by applying for bank loans with individual banks. If successful, the firm implements its current production plan. If it fails, it tries to obtain the remaining external financing it needs from the financial market. This starts the firm's financial market routine.

Financial Market Routine

Here, the firm tries to raise financial capital through the emission of new equity (stocks, bonds) on the primary equity market. If successful, all external financing needs are satisfied and the firm can continue with the actual production routine. Otherwise, the firm needs to revise its production and investment plans. It re-runs the production planning routine subject to the total financial resources available. The new production plan automatically satisfies the budget constraint of the firm, which therefore does not have to run again the financial planning routine. See Fig. 12.2 for a decision flowchart.

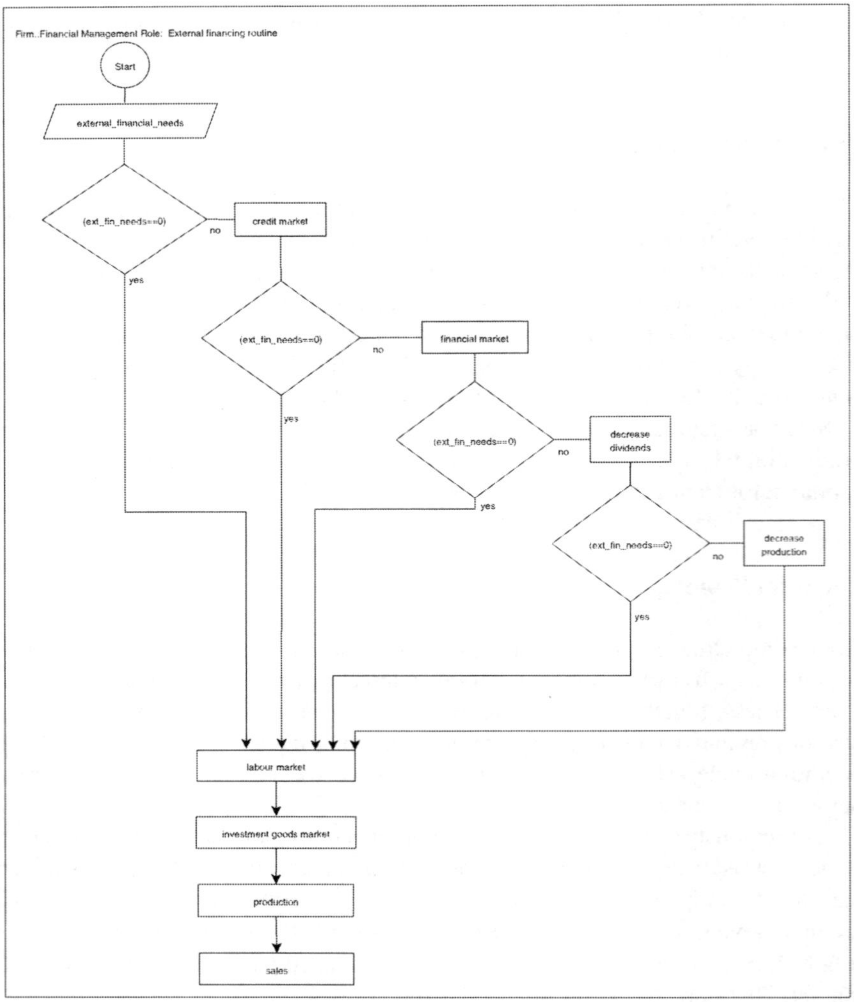

Fig. 12.2 Interface diagram between the firm's financial management role and its credit market and financial market roles. This flowchart shows the conditional functions that need to run to execute the firm's external financing decisions. It first runs the credit market routine and then the financial market routine. If the firm cannot satisfy all its external financing needs, it decreases its dividends (possibly to zero). If it then still cannot satisfy all its liquidity needs, it should re-run the production planning routine to re-optimize the planned production quantity such that the total production costs (i.e., the costs for labour and capital) can be financed

Production Routine

After the production planning routine and the financial planning routine have been completed, the firm starts with the actual production cycle. It first visits the labour market, then the investment goods market, in order to obtain the required factor inputs. At this stage the firm solidifies its production plans, i.e. labour and investments are contracted.

If there is rationing on the labour or investment goods market, the firm spends less than planned and thus has unneeded cash after fulfilling all its contractual obligations. These involuntary savings are deposited in an account at a given bank (different firms can have accounts at different banks). It is deposited there at the end of the firm's operating cycle, i.e., at the end of the day when the firm is active, and will be available for the next production cycle. In that way, all money remains in the system and is available to the banks for making loans to other firms. This prevents inefficiencies due to improductive cash holdings. Incidentally, a similar mechanism is used for modelling the households' left-over consumption budget: all unused cash holdings flow back to the banks immediately in the form of deposits.

The same holds when prices turn out lower than expected: the savings are deposited. If prices turn out higher than expected the firm is rationed in its labour and/or investment demand as well, but there are no savings. The quantity of factor inputs is lower than planned, so also the actual production output will be lower than planned.

After going to the labour market and the investment goods market, the firm starts the actual production process. The output is assumed to be available instantaneously once the financing and factor inputs have been contracted.

The firm distributes the output from a central inventory to the local outlet malls where it tries to sell its merchandise. As previously stated, the selling of output occurs over the course of an entire month, starting from the activation day of the firm. The local outlet malls record and transmit the regional sales revenues. This signals the end of the production cycle, and starts the accounting routine.

Accounting Routine

The firm compiles an income statement and a balance sheet after each production cycle.

The *income statement* includes the monthly cumulative revenues from the sales of goods and services and the total costs: costs of sales, operating expenses, taxes, service of the debt, dividends, etc. (see Table 12.2).

The *balance sheet* lists the total assets and total liabilities (see Table 12.3). The new balance sheet is computed by updating the current cash holdings with the incoming and outgoing cash flows as listed in Table 12.4.

After the balance sheet has been updated, but before executing any payments, the firm first checks if it is financially solvable by checking a bankruptcy condition. If financially solvable, the firm executes its payments and this signals the end of the operating cycle. Otherwise bankruptcy is declared. The precise modelling of events when a bankruptcy occurs is discussed below.

Table 12.2 Firm income statement

Revenues from sales of goods and services
– Cost of sales
– Operating expenses
Earnings before interest and taxes
– Interest payments
Earnings before taxes
– Tax payments
Net earnings
– Debt installment payments
– Total dividend payment
– Share repurchases
Retained earnings

Table 12.3 Firm balance sheet

Assets	Liabilities
Cash holdings	Total debt
Total value physical capital stock	Shareholders' equity
Total value local inventory stocks	

Table 12.4 Firm cash flow

Positive cash flows	Negative cash flows
Revenues from sales	Cost of sales
	Interests
	Tax payments
	Debt installments
	Total dividends
	Share repurchases
Total income	Total costs and expenses

Bankruptcy Conditions

In order to closely mimick managerial practice, the firm checks for bankruptcy at several places during the operating cycle. The first check occurs directly after the balance sheet computation, in order to verify whether the firm can keep its financial commitments from the previous period and pay for the production costs of the upcoming period.

The liquidity needs of the firm are defined as follows:

$$\begin{aligned}
&\text{Financial liquidity needs:} && \text{FLN} = \text{interests} + \text{installments} + \text{taxes},\\
&\text{Production liquidity needs:} && \text{PLN} = \text{labour costs} + \text{investment costs},\\
&\text{Total liquidity needs:} && \text{TLN} = \text{FLN} + \text{PLN} + \text{dividends}.
\end{aligned}$$

Based on these definitions, we define a bankruptcy state, a financial crisis state, and a normal state of affairs:

Cash $<$ FLN:	bankruptcy state
FLN $\leq$ Cash $<$ TLN:	financial crisis state
Cash $\geq$ TLN:	normal state of affairs

In the bankruptcy state, the firm cannot keep its prior financial commitments. Its equity is negative and it has not been able to raise sufficient amounts of financial capital. It sends a bankruptcy message to all the banks with which it has outstanding loans, with content: the credit that is refunded, the bad debt that will not be refunded, and the residual value-at-risk. This information allows the bank to update its balance sheet, which will affect its future lending activities.

In the financial crisis state, the firm did not raise sufficient funds to pay for its total liquidity needs, but can possibly salvage the situation by down-scaling the dividends (in a worst-case scenario, to zero). The down-scaled dividend is given by:

$$\text{Div} = \max\{0, \text{Cash - FLN - PLN}\}. \qquad (12.1)$$

This leads to two possible two sub-states:

1. Cash $\geq$ FLN + PLN and Div ≥ 0.
   The payment account is sufficient to pay for the financial commitments and production costs. In this case the equity of the firm is positive and the financial crisis is resolved. The firm respects its financial commitments and starts the production cycle.
2. Cash $<$ FLN + PLN and Div $= 0$.
   The payment account is sufficient to pay for the financial commitments, but not for the production costs. The equity of the firm remains negative, even after decreasing the dividends to zero. The firm has not yet resolved the financial crisis. It needs to down-scale its production before proceeding further. We assume that the firm pays the financial payments *before* recalculating production costs. Thus, the firm first pays its financial commitments from its current liquidities. It then recalculates the production quantity such that the corresponding production costs are equal to the remaining liquidities:

$$\text{Set prod. costs s.t. PLN} \leq \text{Cash - FLN}. \qquad (12.2)$$

   If the firm succeeds in lowering production costs sufficiently, it has resolved the financial crisis and produces the decreased quantity. In a worst-case scenario, production costs have to be lowered to zero. The firm does not produce anything in the current period.

In the normal state of affairs the equity of the firm is positive. The firm can respect its financial commitments and pay for the production costs as planned. There is no need for a down-scaling of dividends or production. Business as usual continues with the firm entering the labour market and the investment goods market and producing as planned.

12.4 Conclusion

Together with its financial management module, EURACE offers a powerful tool to investigate the yet little understood interaction between real and financial phenomena, including timely questions such as the real impact of credit shortening in a heterogenous, spatially differentiated world. The ability it provides to track the micro-macro interaction and the dynamics into the finest detail may lead to the discovery of transmission mechanisms and phenomena that are obscured by the more traditional representative agent approach. The first simulations we conducted using the financial module indicate that, indeed, the harvest is likely to be rich.

Acknowledgements Research funding by the European Commission as part of the FP6-STREP project EURACE (under contract no. 035086) is gratefully acknowledged.

References

F. Allen and R. Michaely. *Corporate finance: Handbook of the Economics of Finance*, chapter Payout policy, pages 337–429. Amsterdam: North-Holland, 2004.

A. Brav, J. R. Graham, C. R. Harvey, and R. Michaely. Payout policy in the 21st century. *Journal of Financial Economics*, 77(3):483–527, 2005.

S. Coakley, R. Smallwood, and M. Holcombe. From molecules to insect communities - how formal agent-based computational modelling is uncovering new biological facts. *Scientiae Mathematicae Japonicae Online*, e-2006:765–778, 2006.

R. M. Cyert and J. G. March. *A Behavioral Theory of the Firm*. Blackwell, 1963/92.

H. Dawid, S. Gemkow, P. Harting, K. Kabus, M. Neugart, and K. Wersching. Agent-based Models of Skill Dynamics and Innovation. EURACE Report D7.1, Bielefeld University, 2007.

C. Deissenberg, S. van der Hoog, and H. Dawid. EURACE: A Massively Parallel Agent-Based Model of the European Economy. *Applied Mathematics and Computation*, forthcoming.

M. Gallegati, M. Richiardi, and F. Clementi. Agent-Based Models of Goods, Labour and Credit Markets. EURACE Report D5.1. Department of Economics, Università Politecnica delle Marche., 2007.

G. Laycock. *The Theory and Practice of Specification Based Software Testing*. PhD thesis, University of Sheffield, Dept. of Computer Science, 1993.

S. C. Myers. The capital structure puzzle. *Journal of Finance*, 39(3):575–592, 1984.

J. Tirole. *The Theory of Corporate Finance*. Princeton University Press, 2006.

Chapter 13
Serious Games for Economists

Wilbert Grevers and Anne van der Veen

Abstract This chapter explores the methodological issues of modelling human behaviour at the relatively high abstraction level of the differences between individual-based and systems perspectives. Following earlier literature it will be argued that formal frameworks in the social sciences are essentially deductive, as they can be read as implications derived from a set of axioms, or *postulates*. However, in the social sciences implications derived from postulates do not necessarily have a one-to-one correspondence with empirical observations. This marks an important difference with the natural sciences, highlighted by the possibility in the social sciences of choosing one of several formal frameworks. Therefore, a criterion is needed for judging the explanatory value of a model.

An attractive criterion, based on computer simulations using *agent-based models*, is the *possibility of replicating* a stylised version of an observed phenomenon. We argue that parallel to efforts taken in this direction, agent-based methods could also benefit from so-called *serious games*, where techniques from Artificial Intelligence in computer games are adapted for purposes other than entertainment.

13.1 Introduction

The use of formal models in the social sciences naturally focuses on modelling human behaviour. There appears to exist a variety of frameworks for this task and the researcher needs to make a choice. Because nearly all modelling frameworks in the social sciences represent a specific *theory*, different theories can apparently

W. Grevers
University of Groningen, P.O. Box 716, 9700 AS Groningen, The Netherlands, e-mail: `w.a.j.grevers@rug.nl`

A. van der Veen
International Institute for Geo-Information Science and Earth Observation, P.O. Box 6, 7500 AA Enschede, The Netherlands, e-mail: `veen@itc.nl`

be applied for modelling the same phenomenon. This situation characterises the social sciences in general, and it marks a major difference between the social and the natural sciences in connecting models to theories. In the natural sciences, there exists a closer correspondence to an accepted theory and a formalisation that can be used for modelling specific cases.

If the set of models is restricted to mathematical models, the question arises of how a collection of mathematical equations might represent reality. This is easier to answer for the natural than for the social sciences. Especially in the field of applied physics and engineering, the laws that govern the conservation of momentum – Newton's *Laws of Motion* – , the conservation of mass and the conservation of energy are the basic ingredients for nearly any model. For example, the basic set of equations in fluid dynamics, the Navier–Stokes equations, from which many models are derived – ranging from engineering applications to fundamental research –, can be considered a reformulation of basic conservation laws. These conservation laws can be thought of as regularities that can be confirmed empirically. They can be expressed efficiently in mathematical equations that contain variables that correspond to measurable quantities, such as velocity, pressure and mass. In this sense, most engineering models could essentially be considered a restatement of general conservation laws applied in a given context. There exists no equivalence to the conservation laws in the social sciences; there are no 'Fundamental Laws of Human Behaviour'. Perhaps the closest social science ever approached the natural sciences from a methodological perspective is contained in the formalisation of the homo economicus, or *rational actor*. Unfortunately, the success of the rational actor model in representing reality is rather limited in many cases.

Regularities can nevertheless be observed and measured by using statistical methods. Quantification follows in terms of probabilities and frequencies, but statistics does not specify any behavioural rules. If, as in econometrics, probabilities are used for estimating the parameters of a behavioural model, the methodology of the natural sciences seems to be approximated as close as possible. However, a quantification of a 'goodness of fit' is usually needed to verify the validity of the assumptions of the behavioural model. The need for this type of test shows that a social scientist is still confronted with a multitude of models. As a consequence, unlike many natural scientists, the social scientist needs criteria for deciding which model provides the 'best' explanation for an observed phenomenon.

The explanatory value of a theoretical model in the social sciences lies foremost in *deduction*, and will depend entirely on the assumptions from which the implications can be derived. The assumptions therefore have the status of axioms, or *postulates* (Koopmans, 1957), for which no explanations are given in the theory. As a consequence, criteria other than strict empirical verification apply for the evaluation of a deductive theory. In general, a theory is considered to give a good explanation if a minimal number of more or less simple postulates are sufficient to explain relatively complex phenomena. Deductive systems can be formalised by means of mathematical logic. In this sense, mathematical economics – and the use of mathematical models in economic theory in general – can be regarded as the formalisation of deductive reasoning (Koopmans, 1957).

A prominent example is the Arrow–Debreu framework of general equilibrium theory (Arrow and Debreu, 1954; Koopmans, 1957; Debreu, 1959). It is often criticised by both economists and non-economists for its unrealistic assumptions concerning human behaviour. It serves nevertheless as an important reference and benchmark for other frameworks. This applies to economic theory, but also to other social sciences, for example sociology (Coleman, 1990). To organise the discussion of neoclassical economics and its alternatives, following Gintis (2000, p. 43), the label 'neoclassical' will be reserved exclusively for the Walrasian general equilibrium model. In many proposed alternatives, specific elements deviate from the neoclassical assumptions in order to propose a contrasting framework that is intended to yield more realistic results. Frameworks that change elements concerning rationality, while keeping an *individual-based* perspective, in general present a less radical alternative than frameworks that start from a *system* perspective.

The remainder of this chapter is organised as follows. In Sect. 13.2 the characteristics of individual-based methods will be discussed, followed by a discussion of system theories in Sect. 13.3. Section 13.4 is devoted to mathematical biology, as it contains concepts for unifying individual-based and system-based theories, especially in *evolutionary game theory*. Elements from evolutionary game theory are also present in agent-based simulation methods, discussed together with simulation methods in general in Sect. 13.5. In Sect. 13.6, a complementary perspective on agent-based models, based on artificial intelligence in games, is proposed. Section 13.7 concludes this chapter.

13.2 Individual-Based Methods

In the neoclassical general equilibrium model, agents are assumed to make decisions concerning their consumption only on the basis of market prices. This decision is rational in the sense that the agent chooses the consumption bundle he or she prefers over all other bundles. Although this preference structure is usually formalised by a utility function, the decision might in principle alternatively be stated as a set of decision rules. Rather than the decision rules themselves, information about the various options the agent is supposed to have and her or his ability to process information usually pose the actual problem for the modeller. Concerns about the assumed level of rationality often apply to problems about *information*, instead of the ability to make decisions. For the idealised consumer in neoclassical economics all relevant information is contained the known market prices. All interaction between consumers is mediated through these prices.

A much broader set of decision problems is addressed in *game theory* (von Neumann and Morgenstern, 1944). The decisions agents make in game theoretical models are related to strategic interaction. Many types of interaction between human beings – and also between animals – can be thought of as 'strategic' at some relatively general level. As long as at least two agents need to make a decision, and their preference structures are interdependent – meaning that the individual ranking of options by one agent depends on the choice the other agent is expected to make –, there exists a case of strategic interaction. This loose definition of strategic interaction therefore includes nearly every type of human interaction, as long as choices are characterised by dependency and coordination.

The distinction between market interactions and strategic, or *non-market*, interactions is important with respect to the beliefs agents are required to have regarding their environment. Prices can be observed and interpreted by agents with relatively simple cognitive capabilities. In case of strategic interaction, agents will need to have a belief regarding the belief of their opponent. Information on the belief of an opponent is far more difficult to obtain than information on prices. Furthermore, for two agents to reach an agreement, their beliefs regarding the belief of the other agents will need to be consistent. As a consequence, one agent will need to assume that the other agent knows that the first agent knows that he or she knows, etc. This consistency aspect of rationality in game theory may be considered as most problematic. It will be discussed in more detail in Sect. 13.6.

13.3 System Theories

When looking at the mathematical modelling practise, it appears that the label 'systems approach' is usually reserved for systems of differential (and sometimes difference) equations. In some fields of social science, the concept of 'System Dynamics' (Forrester, 1961) seems dominant as the systems approach. System Dynamics can be considered a subset of the all the elements in the more general mathematical theory on dynamical systems. The subset used in System Dynamics, can approximately be identified as the dynamical systems found in *control theory*. Control theory mainly has its applications in electrical and mechanical engineering. A very important concept in control theory is *information feedback*. More elaborate translations from control theory to social science can be found in Cybernetics (Wiener, 1965). A systems approach is also influential in some fields of sociology, following a more qualitative interpretation (Luhmann, 1984). Finally, a systems approach as an interdisciplinary science was founded by Von Bertalanffy in 1951 as the *General Systems Theory* (von Bertalanffy, 1973).

The interpretation of a systems approach remains difficult, even if the systems of differential equations are restricted to systems that bare some resemblance to the description based on control theory. The set of systems of differential equations that contain feedback loops can be made arbitrary large, as any notion of interdependence essentially reflects a feedback loop. The main difficulty with the interpretation of a systems approach in the social sciences is therefore neither its descriptive accuracy nor its lack of it; rather, it still does not supply the researcher with a theory of the behaviour – analogous to 'motion' – of humans. Neither does it provide a theory on the behaviour of systems as other entities or *aggregates* of human beings. It also does not address the issue of aggregation. Therefore, systems theory is not a theory in the usual sense. The theory of dynamical systems is a mathematical theory. As such, it only deals with deductive reasoning not with behaviour.

Notwithstanding the difficult interpretation of a systems approach as a general theory, the results from research on *complex systems* – as far as it is restricted to the mathematical theory of connected dynamical systems – have had implications for other theories as well. The combination of feedback mechanisms and non-linear differential equations gives rise to several complex phenomena. A system perspective is able to accommodate many features of the so-called *complex systems theory*. Furthermore, 'complexity' is often identified with *interacting individuals*, which does not exclude an individual-based approach, or methodological individualism, as in economics (Anderson and Arrow, 1988; Arthur et al., 1997; Blume and Durlauf, 2006). While some success has been achieved in the operationalisation of non-linear systems in explaining regularities in physics, the most successful discipline in this respect is arguably biology. Perhaps because biology also deals with behaviour, some economists seem to follow ideas first developed in *mathematical biology* (Murray, 1990) and *mathematical ecology* (May, 1974) for introducing elements from complex dynamical systems in economic theory.

13.4 Mathematical Biology and Game Theory

Lewontin (1961) was one of the biologists that introduced game theory for analysing population genetic theory, extending the seminal work on the statistical approach to population genetics by Fisher (1930). Applications of concepts from game theory to the analysis of conflict between *behavioural types* as fractions within a population followed (Maynard Smith, 1974; Taylor and Jonker, 1978). Instead of the direct interaction that consists of eating and being eaten in the Lotka–Volterra system, these models were based on the evolutionary advantage of adopting a certain behavioural strategy. Within both biology and economics, the analysis of these so-called *population games* is referred to as evolutionary game theory (EGT) (Hofbauer and Sigmund, 1988; Weibull, 1995; Hofbauer and Sigmund, 1998).

Population games seem to offer a solution for the both the system interpretation of complex dynamical systems and the interpretation of individual beliefs in decision making. On one hand, dynamical systems in evolutionary game theory have a clear reference to individuals in a population. The interpretation of a fraction of a population only has a strict mathematical justification in the assumption that the population is very large, but as an approximation this interpretation is often valid more generally. The notion of evolution, on the other hand, offers an interesting alternative to the consistent belief systems of interacting agents. Admittedly, this notion is relatively abstract, but for example the evolution of institutions and conventions might be interpreted as the result of a repeated game, where people have adjusted their strategies over time (Young, 1998). Whether people in a country drive on the right or left-hand side of the road is not important; most important is the agreement on one of the two sides. Although, the agreement can be interpreted in terms of strategic interactions, the problem of consistency in the beliefs of all agents on the road is delegated to legislation that has evolved over time. It is therefore only rational for the individual agent to obey the current convention and no extreme cognitive capabilities will need to be assumed.

13.5 Simulation Methods

The word 'simulation' carries several meanings in relation to modelling. A common element is the reliance on a *computer* for model implementation. The more conservative meaning follows physics and engineering where simulation is often identical to numerical computation. Given the extensive use of differential equations in the natural and engineering sciences, often both simulation and numerical computation can read as synonyms for *numerical integration* of mathematical models of dynamical systems. Simulation – by means of direct integration – in case of a dynamical system, usually concerns the reconstruction of a position or volume of an entity, given a description of its velocity or growth rate, using small but finite approximations for infinitesimal small quantities in the analytical equivalents. The use of a computer becomes necessary if the system of equation cannot be solved analytically, deriving a closed form solution with pencil and paper. In the natural sciences analytical solutions are often preferred, as they suggest a more precise solution. Analytical solutions are also usually preferred in economics, though for a slightly different reason. In economic theory an analytical solution conforms to a deductive *proof*, whereas a numerical solution can only represent a special case.

A different approach to numerical integration is presented by *Monte Carlo* simulation or integration methods (Judd, 1998), often applied in econometrics. Simulation in econometrics has a slightly different meaning. There, it mainly concerns the estimation of the mean of a variable, relying on integration of this variable times a probability density, by calculating the average value on the basis of a very large number of draws from the probability density distribution.

Similar to evolutionary game theory is a specific modelling approach in which models are formalised directly in computer code. Once the decision is made to model system phenomena only through a representation of the behaviour at the individual level, a next step might consist of taking advantage of an object-oriented computer language. This approach adopts concepts also applied in computer science (Weiss, 1999; Wooldridge, 2002). Although these so-called *agent-based models* can in principle be programmed in a procedural language as well – executing one command or function after another – the main benefit of an object-oriented (OO) language comes especially with the concept of *encapsulation*. In any OO language the focus is on objects in which data are stored. These data are often called attributes. If, for example, a data object represents an employee, one of his attributes could be his wage. The concept of encapsulation only refers to the possibility for the programmer to store the attribute in the object so that it is 'hidden', in order to force the programmer to adopt the discipline of defining functions – or 'methods' – for accessing the attributes. These methods are interesting for modelling human behaviour, because they relate information to communication. In this way, agents will only reveal their information when asked to do so explicitly. Beliefs can thereby become consistent in a more convincing way, since they can be reconstructed with an elementary communication protocol. Agents can only become aware of the information other agents posses if they are able to 'ask in person'.

The idea of convincingly reconstructing a certain result lies at the heart of Agent-based Computational Economics (ACE). Epstein (2007, p. 8) even explores the implication of posing it as a requirement:

> If you didn't grow it, you didn't explain its emergence.

Interestingly, similar considerations have been discussed within the context of using computational methods in economic in general, not only restricted to its agent-based variant, by Judd (1998). Judd (1998, p. 13) takes a more or less defensive position when arguing that numerical results might serve as a solution good enough for decision-makers that are likely not to be interested in the proof of a theorem. The reference to a theorem shows that Judd interprets computational results as an approximation to the deduction of a theorem, in line with Koopman's (1957) interpretation of neoclassical economics as an axiomatic theory. Axelrod and Tesfatsion (2006, p. 1650) expand this theme when it concerns agent-based computational economics, in their 'third way of doing science'. Nevertheless, also there the main application of the agent-based model is to the ability to deductively generate *synthetic data*.

13.6 AI in Computer Games

As was discussed in Sect. 13.5, one approach to agent-based modelling consists of replicating stylised facts with a large number of agents, where each agent is equiped with a set of relatively simple behavioural rules. This approach is related to the use of methods from statistical physics (Durlauf, 1997), which – in turn – are closely related to methods in evolutionary game theory. An alternative approach might be offered by methods from computer science, especially *Artificial Intelligence* (AI) in games (Millington, 2006). The reliance on agents and OO implementation remains similar, but the cognitive capabilities of the agents will need to be more advanced, possibly reaching as far as the *cognitive architectures* (Newell, 1990), such as *Soar* (Laird et al., 1987). The level of cognitive sophistication is, however, likely to depend on the application. The implementation of AI in computer games is often rather pragmatic, using combinations of various techniques. A suitable reference point for exploring possible applications of similar techniques in the realm of economics is offered in similar games developed for use outside the entertainment industry; the so-called *serious games*.

From a scientific perspective, adopting game AI in economic models might imply a shift from understanding human behaviour to *exploring possible strategies* in a given context. The formalisation of many aspects of economics in terms of game theory (Vega-Redondo, 2003) offers a natural benchmark for building agents that have 'winning the game' as their goal. The modelling exercise would in that case need to focus primarily on the formalisation of a real problem as a game, similar to, for example, a board game. Next, rules addressing the way the game is played need to be defined. The most famous example of this approach is Axelrod's tournament (Axelrod, 1984). More recent advances in designing agents for tournaments seem to take place mainly within computer science, relatively isolated from the social sciences (Wellman et al., 2007).

Along with the formalisation of competition in economic theory using concepts from game theory, the Nash equilibrium (Nash, 1950) – as the dominant solution concept – can serve as the main link between AI techniques in games and economic models. The informal definition of the Nash equilibrium maintains that the individual agent chooses a strategy according to his *best response to the best responses* of all other agents. It can therefore be considered as a more specific instance of the consistency issue discussed in Sect. 13.2. It is very demanding for the cognitive capabilities of all the agents involved, as they are required to have *rational expectations* (Muth, 1961) while optimising. Not only do they need to be aware of their own best response, but also of the best responses of all other agents. Since the other agents, in turn, are required to play their best response to the best response of the first agent, all agents should play the strategy that confirms to the expectations of all other agents. The Nash equilibrium in non-cooperative games can therefore also be interpreted as the requirement for the beliefs of agents involved to be consistent in self-interested competition.

The problems with the interpretation of the Nash equilibrium as a solution concept are well-documented (Gintis, 2000; Vega-Redondo, 2003; Bowles, 2004). One issue that arises, is the need for agents to possess a complete insight in the preferences of all other agents, in order to determine their best responses. Another issue concerns the computational complexity of the Nash equilibrium. Recent mathematical proofs show that even relatively simple Nash equilibria belong to a class that is not computable in polynomial time (Chen and Deng, 2006). Hence, even if all information necessary would be available to compute the Nash equilibrium, there exists no efficient algorithm for finding an exact solution.

In the analogy with games played by individuals for entertainment, the Nash equilibrium as a solution concept can also be considered problematic for a more pragmatic reason. A best response to best response resembles the definition of a *tie* in many competitive games. The goal for players in a competitive game, however, is to *win*, rather than reaching a tie. In any tactical game with two players, such as chess, one way of trying to win is to speculate on the opponent making a mistake. Referring to the rational expectations interpretation of the Nash equilibrium in a heuristic way, one could suggest that the beliefs of the winning player in fact were consistent, anticipating the mistake of the opponent. By the same argument, the opponent must have had *inconsistent* beliefs, as he did not anticipate the first player's strategy, implying that he expected him to do something else. It might be argued that this analogy applies primarily to zero-sum games, while many games that are of interest to economists are non-zero sum. However, for the given heuristic context, this difference is less strict than in mathematical game theory. For example, in some game played purely for entertainment by two agents, the goal game could be to collect points. One of the rules of that game might state that the player who has earned most points after ten rounds is the winner. According to this rule, there can only be one winner, unless there is a tie, hence in this sense the game is zero-sum. Nevertheless, the both agents can earn points. In a similar way, two producers could compete for market shares. The producer with the largest share can be considered the 'winner', but as long as the other producer can survive on his share, both producers can coexist on the market. Extending this last example, if the two producers are identical, the economic theory of imperfectly competitive markets would often reside to a Nash equilibrium, implying that the market shares for both producers would be equally large. Instead of a focus on the question how this equilibrium can be reached, a game AI approach to this model would stress the competitive aspect by trying to develop an agent that employs strategies for achieving a larger market share than the opponent.

With a strong focus on the competitive element in economic games, there is admittedly a risk that insufficient attention will be paid to motivations of real economic agents, other than 'defeating' opponents. It sustains the caricature of the self-interested individual in economic theory, neglecting altruistic behaviour (Bowles, 2004). However, it might be argued that the introduction of different motivations would alter the rules of game. Different priorities on the behalf of the agents, could be formalised as a subset of the rules that is strictly personal. In the extreme case, general rules would only determine the minimal conditions within which agents could try achieve their individual goals. This would result in a different type of model. From a methodological perspective, a competitive game with a universal set of rules has the advantage that is allows for a strict separation of rules and behaviour. The rules define the core of the model, while the behavioural aspects are covered by the design of the agents. Instead of defining universal laws of human behaviour, the modeller can impose by the formalisation of an economic problem *as* a competitive game the main simplification of the real situation. Although in reality not every producer might be maximising profit, the formalisation as a game in which profit maximisation is part of the rules, might in many cases present a useful stylised representation. Next, this model would require agents that populate it. These agents do not need to follow universal rules either, as long as they are designed to win the game, simulation runs with several combinations of different agents can give an indication of how competition might evolve. This approach allows for a variety of methodologies to the design of agents, similar to AI applications in games.

The goal of AI applications in video games can be summarised as presenting a credible human-like opponent for a human player. Although this goal itself is rather ambitious, the need for practical applications urges developers to be pragmatic in finding solutions. A wide range of techniques that are part of the academic literature on AI and computational techniques are present in the literature on AI in games as well. This includes methods from computational intelligence, such as artificial neural networks, genetic algorithms, and fuzzy logic (Millington, 2006).

Often, algorithms employed need to be very efficient, as the computer power needs to be shared with the graphics and other elements of the game. Besides that, the search for credible opponents is distinct from the goal of *replicating* the behaviour of real humans. With respect to economic modelling, if the main part of the model consists of the formalisation as game, computer agents need not to behave as real humans in all respects, as long as their approach to the game can render some insight to structure of the game.

Nevertheless, in some respects the AI in entertainment games can fail short for agent-based computational economics. Because the credibility of a computer opponent primarily depends on the entertainment value of the human player – who, as a customer, bought the game – cheating is sometimes allowed. Since the computer opponent usually is an integral part of the game, the developer can introduce short cuts in the information processing capabilities. In a shooting game, for example, the computer opponent can be made aware where the human player resides on the map, without the need to be able to realistically see him, facilitating chasing behaviour. Although this might enhance the game-play, in terms of speed, etc., agents in economic models of the type discussed in this chapter are supposed to specifically address issues of information the assumption of rational expectations avoids.

Examples of AI in games that reach beyond the purpose of entertainment only can be found in so-called *serious games*. Serious games apply elements usually found in games in virtual environments for training or education. Part of the development of serious games is driven by the demand for military training tools, especially urban combat. Another current field of application is health care, both for training medical personnel and therapy for patients. Closer to economic modelling are business simulation games. Traditionally, simulation methods already have a strong position in business administration and operations research. In this respect, AI in serious games might serve as a common language for agent-based models in economic theory.

13.7 Conclusions

In this chapter, a general overview is presented of the two main methodological approaches applied in the social sciences. The first is the systems approach, the second the individual-based approach. Both are introduced after highlighting the contrast between the natural and the social sciences. The possibility for a researcher to choose a specific methodology when building a model in the social sciences is identified with the absence of general laws. This absence especially has implications for the deductive nature of model building in general. Deductive models in the natural and engineering sciences most frequently can be considered restatements of fundamental conservation laws, for example, the conservation of mass or energy. In an applied model these laws themselves are not verified. As a result, the axiomatic nature of the fundamental laws use can be accepted universally by the researchers in the discipline.

Deductive models in the social sciences rely on axioms as explicit assumptions or postulates. As a consequence, the use of these models always consists of a thought experiment. In this respect, microeconomics adopts an individual-based approach, based on certain postulates. Although these postulates are problematic in several cases, in this thesis the formulation of postulates at the level of individuals is preferred over assumptions regarding the behaviour at the level of aggregates or systems. A systems approach frequently relies on mathematical control theory. From a methodological point of view, this is problematic, as the issues of behaviour and aggregation are not addressed. Nevertheless, the mathematics of complex dynamical systems can offer the possibility to assess self-organisation at the system level, if the behaviour of the aggregate can be related to the behaviour of interacting individuals. Examples of this approach can be found in mathematical biology, especially in evolutionary game theory (EGT).

The relation between simulation and computation might be closer than is sometimes claimed in the literature on the use of simulation in the social sciences. An interesting approach is suggested in the discipline of agent-based computational economics (ACE), where some researchers propose to replace a mathematical proof by a credible simulation run.

Finally, it is argued that while ACE often relies on simulations with a large number of agents with relatively simple rules, agent-based approaches in economics might also benefit from techniques from Artificial Intelligence developed for applications in games. So-called *serious games* in business applications could offer a common language, possibly adopting cognitive modelling techniques, such as *cognitive architectures*. This, however, implies a different approach to economic modelling, with more emphasis on defining the rules of the game, while behavioural models are used for exploring ways the game might be played.

Acknowledgements The authors would like to acknowledge the Dutch Knowledge Network on System Innovations and Transitions (KSI) for the financial contribution during the initial phase of this study. Wilbert Grevers would also like to thank the NWO (Netherlands Organisation for Scientific Research)-ACTS *sustainable hydrogen* programme for financial support during the final phase.

References

P. W. Anderson and K. J. Arrow, editors. *The economy as an evolving complex system*. Addison-Wesley, Redwood City, CA, 1988.

K. J. Arrow and G. Debreu. Existence of equilibrium for a competitive economy. *Econometrica*, 22:265–290, 1954.

W. B. Arthur, S. N. Durlauf, and D. A. Lane, editors. *The economy as an evolving complex system II*. Addison-Wesley, Reading, Mass, 1997.

R. Axelrod. *The Evolution of Cooperation*. Basic Books, New York, 1984.

R. Axelrod and L. Tesfatsion. A guide for newcomers to agent-based modeling in the social sciences. In Leigh Tesfatsion and Kenneth L. Judd, editors, *Handbook of computational economics: agent-based computational economics*, volume 2. Elsevier, Amsterdam, 2006. appendix.

L. E. Blume and S. N. Durlauf, editors. *The economy as an evolving complex system III*. Oxford University Press, Oxford, 2006.

S. Bowles. *Microeconomics: behavior, institutions, and evolution*. Princeton University Press, Princeton, 2004.

X. Chen and X. Deng. Settling the complexity of 2-player nash-equilibrium. In *Proceedings of the 47th Symposium on Foundations of Computer Science*. IEEE Press, 2006.

J. S. Coleman. *Foundations of Social Theory*. Belknap Press of Harvard University Press, Cambridge, MA, 1990.

G. Debreu. *Theory of value: an axiomatic analysis of economic equilibrum*. Wiley, New York, 1959.

S. N. Durlauf. Statistical mechanics approaches to socioeconomic behavior. In W. B. Arthur, S. N. Durlauf, and D. A. Lane, editors, *The Economy as an Evolving Complex System II*. Addison-Wesley, Reading, Mass, 1997.

J. M. Epstein. *Generative Social Science: Studies in Agent-Based Computational Modeling*. Princeton University Press, Princeton, 2007.

R. A. Fisher. *The genetical theory and natural selection*. Clarendon Press, London, 1930.

J. W. Forrester. *Industrial Dynamics*. Productivity Press, Cambridge, 1961.

H. Gintis. *Game Theory Evolving*. Princeton University Press, Princeton, 2000.

J. Hofbauer and K. Sigmund. *The theory of evolution and dynamical systems: mathematical aspects of selection*. Cambridge University Press, Cambridge, 1988.

J. Hofbauer and K. Sigmund. *Evolutionary games and population dynamics*. Cambridge University Press, Cambridge, 1998.

K. L. Judd. *Numerical methods in economics*. MIT, Cambridge, MA, 1998.

T. C. Koopmans. *Three Essays on the State of Economic Science*. McGraw-Hill, New York, 1957.

J. E. Laird, A. Newell, and P. S. Rosenbloom. Soar: An architecture for general intelligence. *Artificial Intelligence*, 33(1):1–64, 1987.

R. C. Lewontin. Evolution and the theory of games. *Journal of Theoretical Biology*, 1:382–403, 1961.

N. Luhmann. *Soziale Systeme*. Suhrkamp, Frankfurt am Main, 1984.

R. M. May. *Stability and complexity in model ecosystems*. Princeton University Press, Princeton, NJ, 2nd edition, 1974.

J. Maynard Smith. The theory of games and the evolution of animal conflic. *Journal of Theoretical Biology*, 47:209–221, 1974.

I. Millington. *Artificial Intelligence for Games*. Elsevier, Amsterdam, 2006.

J.D. Murray. *Mathematical biology*. Springer, Berlin, 1990. Corrected 2nd print.

J. F. Muth. Rational expectations and the theory of price movements. *Econometrica*, 29(3):315–335, 1961.

J. F. Nash. *Non-cooperative Games*. PhD thesis, Princeton University, 1950.

A. Newell. *Unified theories of cognition*. Harvard University Press, Cambridge, MA, 1990.

P. D. Taylor and L. Jonker. Evolutionarily stable strategies and game dynamics. *Mathematical Biosciences*, 40:145–156, 1978.

F. Vega-Redondo. *Economics and the theory of games*. Cambridge University Press, Cambridge, 2003.

L. von Bertalanffy. *General System Theory: Foundations, Development, and Applications*. Penguin, London, 1973. (first published in 1968).

J. von Neumann and O. Morgenstern. *Theory of Games and Economic Behavior*. Princeton University Press, Princeton, 1944.

J. Weibull. *Evolutionary Game Theory*. MIT Press, Cambridge, MA, 1995.

G. Weiss, editor. *Multiagent systems: a modern approach to distributed artificial intelligence*. MIT Press, Cambridge, MA, 1999.

M. P. Wellman, A. Greenwald, and P. Stone. *Autonomous Bidding Agents*. MIT Press, Cambridge, MA, 2007.

N. Wiener. *Cybernetics or control and communication in the animal and the machine*. MIT Press, Cambridge, MA, 2nd edition, 1965.

M. J. Wooldridge. *An introduction to multi-agent systems*. Wiley, Chichester, 2002.

H. P. Young. *Individual Strategy and Social Structure: An Evolutionary Theory of Institutions*. Princeton University Press, Princeton, 1998.

Part V
Invited Speakers

Chapter 14
Computational Evolution

Peter A. Henning

Abstract The paper presents an introduction to simulation models for evolutionary systems from three different points of view. Macro evolution is discussed in view of catastrophic events, like e.g. extinction events – and it is shown that these may occur at random, without external cause. Micro evolution is demonstrated looking at an actual physical problem. It turns out that there is an important distinction between natural evolution and the usage of evolutionary models for optimization purpose. In the third section we consider bottom-up evolution, by discussing a recent model for the generation of simplified discrete models for large arrays of regulatory circuits – to the end of creating artificial life. The author takes the liberty to mention links to economical systems whenever possible.

14.1 Introduction

When looking at their field of interest, economists usually have a vague concept of dealing with something "evolutionary." Clearly, on a superficial level this is true: The system is complex, its constituents are living entities and, well, some evolution undoubtedly takes place. Also on a much deeper mathematical level similarities exist: Economical as well as ecological systems are exhibiting *complex* behaviour that somehow emerges from the interaction of much simpler, even *simple* constituents. Consider for example the catastrophic failure of species in an ecosystem – is it not very similar to the catastrophic failure of enterprises in a dynamical market? However, apart from such "we-feel-there-must-be-some-similarity" explanations, not many bridges have been drawn from evolution into economy and vice versa.

At least one of these existing bridges moreover is known to be ill-founded: Many of us still feel, that evolution leads to something better in due time – e.g. as the

P.A. Henning
Institute for Computers in Education, Karlsruhe University of Applied Sciences, Moltkestrasse 30, 76133 Karlsruhe, Germany, e-mail: peter.henning@ice-karlsruhe.de

"clever" homo sapiens has evolved from the "stupid" ape. Successful economical projects therefore are sometimes thought to be the better animal in a world subject to a paradigm of "survival of the fittest." This however is a severe misconception of evolution, because evolution does *not* lead to a systematic improvement, as will be shown in the following.

It is also beyond doubt that the interdisciplinary view on evolution and economy might have its limits. Understanding complexity in one system, plugging the results of this understanding into a simulation of another system and finally producing pictures that are similar to reality does not imply understanding of the second system. At best, one may argue that the two systems belong to a same class in behaviour. This was shown a long time ago for physical systems exhibiting phase transitions, where the behaviour classes are called *universality classes*.

While it is an appalling idea to look for universality classes also in economical and ecological systems, such a task would be much beyond the scope of the present paper. Rather, it is restricted to give an overview about foundations and recent developments in the informational understanding of evolution, e.g. into the question of how complexity emerges from simpler constituents. We will therefore discuss several simple models from computational evolution and point out their possible relevance for economical systems.

The genetic constituents of individuals are yet another matter. How does, e.g. one obtain the look and feel of an object called a "cat" from the interaction of chemical substances? Can one, in other words, hope to understand the emergence of complexity also on this level? One of the developments discussed below should be able so set a foundation for the formidable task of artificial life.

In the following, we are therefore dealing with a multi-scale view of evolution: an ecosystem, its species, their individuals, cells and also genetic constituents. This multi-scale view on evolution has been proposed by Gould (2002).

Individuals are organized in groups, called populations – and a population is the smallest concrete instance of a species. From an evolutionary viewpoint, single individuals are not important – in the language of computational physics, they are only "test particles" set on a random track governed by interaction with other test particles. Consequently, if removing any single individual from a population this will not change the evolutionary path of the species. Within a population, individuals may exchange genetic material, and hence a population may be seen as genetically homogeneous.

In our simplified view, genetic material is not exchanged across population boundaries. Consequently, if a species consists of isolated populations, these may evolve differently – thereby creating new species. Note, that such an isolation may be due to geographical separation (allopatric resp. peripatric speciation), but also due to other separating influences (parapatric resp. the – still controversial – sympatric speciation) (Baker, 2005; Rice and Hostert, 1993). Furthermore we have to keep in mind, that due to modern biological knowledge, genetic material may be exchanged even across species boundaries – consequently, reality is much more complex than our simple model even from the start.

Within these limits, we now discuss several evolutionary models with their calculational details and their possible relevance for economists. These are, in particular

Macro evolution	Complex behavior of the complete ecosystem. Species are developing, even mutating – but their details, or rather their capabilities are not of interest.
Micro evolution	Complexity emerging in the blueprint of a species. Species are developing and mutating – and of interest are the varying capabilities of each species and their expression in the interaction with other individuals.
Bottom-Up evolution	Complexity in the behavior of a larger system (a group of genes, cells, individuals, or an ecosystem) arising from capabilities of its building blocks.

An important question in this context is the relationship between micro and macro evolution. Does micro evolution automatically lead to macro evolution? Is it a sufficient or necessary condition for the latter? It should be kept in mind, that the notion of 'automatic' macro evolution is very much to the dislike of creationists, because it is a clear argument against the need for a creator.

14.2 Catastrophic Events in Macro Evolution

One problem that baffles economical as well as evolutionary scientists is the occurrence of catastrophic events – like stock market crashes and extinction events. This is of course due to the fact that we tend to think linearly: Big events, at the first look, need big causes – and having identified a big cause for catastrophic events usually is considered as an *explanation*. Consequently economists even today are searching to explain the New York stock market crash of 1929 (Eichengreen and James, 1996; Galbraith, 1997), where a definite cause has not been identified. Palaeographers on the other hand are searching for causes of historical extinction events.

In view of this, the Alvarez hypothesis of an asteroid hitting earth and thereby creating the Cretaceous–Tertiary (K-T) event of dinosaur extinction, has a large predictive power and is not seriously doubted nowadays. But, unconsidered in most cases, we know that many more extinction events have happened in history. Table 14.1 shows some of other the major events known in history – many more, however, are known (Bambach et al., 2004). Indeed, a closer analysis of major and *minor* extinction events throughout history reveals, that the probability $H(s)$ to find an exctinction event where s percent of all species have become extinct varies according to a power law as

$$P(s) \propto s^{\eta} \tag{14.1}$$

with a value of $\eta = -2.3$. So far, research for causal effects behind these extinction events has not revealed anything. Of course, many speculations are around, ranging from volcanic activity and methane eruptions to cyclic disturbances of the

Table 14.1 Large extinction events in history, given in Millions of years (Ma) before our time

Paleographic name	Age (Ma)	Recovery time (Ma)	Effect	Cause
Cretaceous-Tertiary	65	20	Extinction of 60% of all animal species	Asteroid Impact
Trias	210	100[a]	Extinction of 35% of all families	?
Perm	250	100[a]	Extinction of 50% of all animal families, 95% of all marine species	?
Devon	360	30	Extinction of 30% of all animal families	?
Ordovicium	440	25	Extinction of 50% of all animal families	?

[a] Together

Oort cloud and subsequent cometary bombardment of the inner solar system. In the recent past, the asteroid impact causing the K-T event has been traced to a collision in the asteroid belt – while the concept of a cyclic disturbance, originally finding a 26 Ma period, nowadays has been revived as having a 62 Ma period (Muller and Rohde, 2005).

However, from the standpoint of nonlinear science, these searches for big causes behind big events are unnecessary. In the following we will demonstrate this by a very simple model of macro evolution, originally due to Solé and Manrubia (1996). We start by considering an ecological system Ω consisting of N species, each occupying its own ecological niche. None of the details of a real ecological system is of interest here – we restrict the knowledge of the system at time t to a single state bit S_i for each species i

$$\Omega_t = (S_1, S_2, \dots S_N)^T \, , \, S_i = \begin{cases} 1 \text{ if species } i \text{ alive} \\ 0 \text{ if species } i \text{ extinct} \end{cases} \tag{14.2}$$

where the T indicates transposition, i.e. Ω is a column vector. At the beginning, each species is alive, e.g. all the ecological niches of the system are occupied. The initial state vector of the ecosystem therefore is

$$\Omega_0 = (1, 1, \dots 1)^T. \tag{14.3}$$

We now introduce a discrete temporal evolution pattern: The state of each species i at time $t+1$ is derived from its ancestor at time t by acting on it with a nonlinear function. This function is constructed by weighing the influence of all other species on species $j \neq i$ with a positive or negative value. Formally this expressed by a time dependent $N \times N$-matrix M_t called the connection matrix,

$$M_t = (m_{ij}) \, , \, i, j = 1, \dots N \tag{14.4}$$

with has elements in the range $m_{ij} \in [-1, 1]$ and starts its evolution as a diagonal matrix.

$$M_0 = \text{diag}(1, \dots 1). \tag{14.5}$$

The multiplication of the bit vector Ω with M maps it to a real number vector

$$\omega = (s_1, \ldots s_N)^T \tag{14.6}$$

with elements comprising the sum of all weighted influences

$$s_i = \sum_{j=1}^{N} m_{ij} S_j. \tag{14.7}$$

This value is either positive – in which case the species remains unaffected. Or it is negative, in this case the species becomes extinct. The time evolution of the ecological system is the described as a two-step process

$$\begin{aligned} \omega_{t+1} &= M_t \Omega_t \\ \Omega_{t+1} &= H(\omega_{t+1}), \end{aligned} \tag{14.8}$$

where H is a vectorized Heaviside step function, acting on each element s_i of the real number vector ω by

$$H(s_i) = \begin{cases} 1 \text{ if } s_i > 0 \\ 0 \text{ if } s_i \leq 0. \end{cases} \tag{14.9}$$

The ecosystem is in stationary equilibrium unless some random factor is introduced by replacing matrix elements of M with random numbers from the interval $[-1, 1]$. We find it realistic to assume, that in each time step only N matrix elements may change. To be more specific: For each $i \in [1, N]$ we calculate a random integer $k \in [1, N]$ and a random real number $\mu \in [-1, 1]$ and then set

$$m_{ik} = \mu. \tag{14.10}$$

Note, that such a step is not necessarily a mutation. If species i within a generation starts to "eat" species j instead of cooperating with it, this might be a learned response due to environmental pressure.

Furthermore, speciation is included in the model: All the ecological niches that are vacated by extinction of species are reoccupied by this speciation process. For simplicity we assume, that all vacant niches with $s_i = 0$ are occupied by speciation of a single, randomly selected surviving species l. We may safely assume, that in this process also the connection matrix elements of the "branching" species are copied – up to a random small variation (say, of $\pm 1\%$). Note, that this behavior – i.e. the branching of a single species into many empty niches – is indeed a finding of palaeography. For a concrete example we run a simulation with $N = 250$ for 160.000 generations (of which the first 10.000 are discarded as giving transient results). The resulting extinction rate is displayed in Fig. 14.1. In 70.5% of all generations not a single species becomes extinct – but we also find rare catastrophic events where 246 of 250 species suddenly die out. Indeed, fitting the power law of (14.1) to the result, we obtain that in this example the exponent takes a value of $\eta = -2.26$ (see

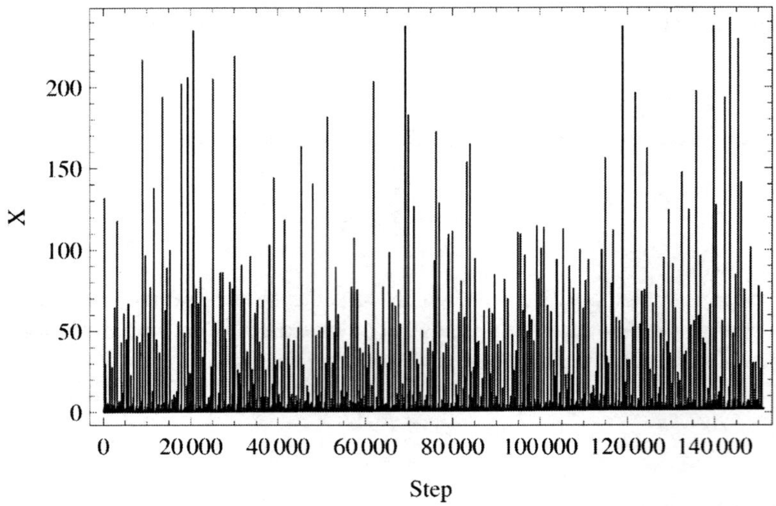

Fig. 14.1 The number of extinct species X as a function of the generation number, i.e. the extinction rate

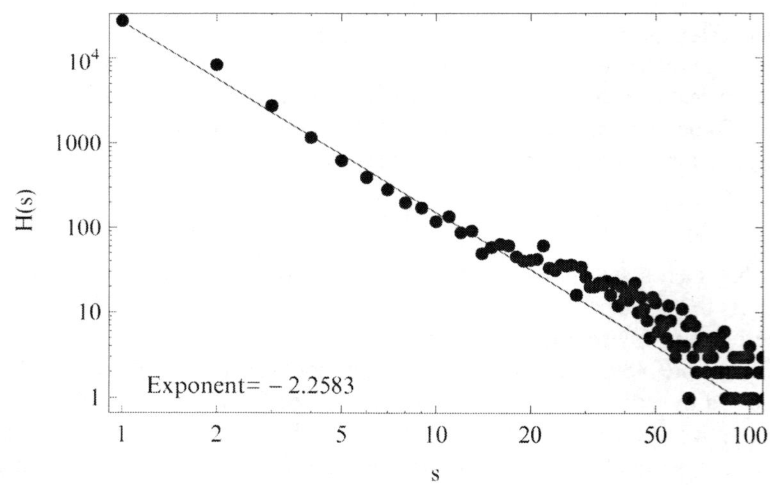

Fig. 14.2 The probability for an extinction event where s species die out and its comparison to a power law, (14.1)

Fig. 14.2). Of course, the similarity of this value to the one fitting the palaeographic results and quoted after (14.1) is accidental.

Another lesson to be learned from the data is that it is free of any time scale. This may be seen by plotting the distribution of temporal (or rather generational) distance of events as a function of the percentage of extinct species, see Fig. 14.3. We therefore draw the following conclusions:

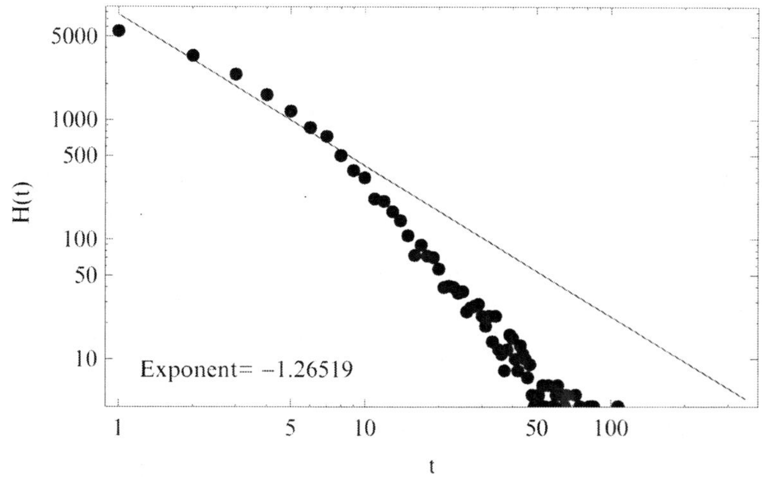

Fig. 14.3 The probability to have t generations without extinction and its comparison to a power law

1. Obviously, a temporal structure of smooth development punctuated by catastrophic events does not require large (or even individual) causes for these catastrophic events – with striking conjectures on natural evolution (Eldredge and Gould, 1972). Conversely, catastrophes occur "by themselves" in a purely random fashion simply from the dynamics of the system.
2. The agents responsible for the dynamical behavior may be very simple, their inner micro-evolution is completely irrelevant for large scale dynamics as long as it changes their interaction with the environment.
3. Macro evolution therefore has the necessary precondition of collective behavioral change – which is a much weaker requirement than micro evolution. The findings of the Solé-Manrubia model therefore lend strong evidence to the arguments of Gould (2002) who proposed a hierarchical concept of evolution, where one level of hierarchy is not necessarily based upon the other.

Another interesting view on this model results from looking at the state vector Ω_t. The evolution process may be described as acting on a finite string of bits with a function – and replacing it with another bit string. Such a mapping is called a "cellular automaton" (Ermentrout and Edelstein-Keshet, 1993) – and cellular automata have been proven very useful in multi-scale modeling of systems.

14.3 Variations of Micro Evolution

In micro evolution, we are not really interested in the large scale behavior of the system, but in the capabilities of individuals. A fully developed ecosystem may consist of thousands to millions of species, and species in reality may have billions to trillions of individuals. We are therefore now considering systems that are several, if not many orders of magnitude smaller than an ecosystem. Macro evolution generally runs too slow to be noticed by us, and molecular evolution runs much faster than expected in ordinary life. The intermediate micro level grants us a more manageable view: Micro evolution may even be "witnessed," e.g. in breeding new plants. While we want to keep this notion of "improval" as a tentative hypothesis for (computational, or artificial) micro evolution, it has been stated already in the introduction that this requires some careful analysis.

Hence, for a starter we are looking at evolutionary models for the *improval* of concepts (genotypes), their realization (phenotypes), strategies (behavioral patterns) or simply ideas. Suppose we have a group of individuals with a variety of capabilities – some of them are "better" (those we want to keep, to breed, i.e. to develop) and others are "worse." These latter individuals are considered unwanted. We then have the hope that all "unwanted" individuals and species die out – if not rapidly, at least after some short time.

However, here is a catch: It must be stated clearly that this notion of *immediate or rapid* removal of unwanted individuals is an idealized belief at best – because it requires the presence of an infinite or very large sink of entropy, or source of information (the detailed analysis combines non-equilibrium thermodynamics with information theory, which is beyond the scope of the present paper). If such an information source is not present, or if it may be exhausted, "unwanted" individuals may cloud the gene pool for many generations. A nice phenomenological example for this limitation is the development of an ideal predator species, which may have a stable population only in case the supply with "victims" is unlimited.

As capabilities of individuals we can either view the genotype, i.e. the formal description of an individual, or its phenotype, i.e. a real world realization of the genotype. In this latter case, one has to keep in mind the additional difficulty of relating genotype and phenotype: Is there a unique relationship, or can a single phenotype also be associated with more than one genotype? Such might be the case when the genotype is a construction plan, because there might be more than one way to build a technical gadget by interpreting the plan. Depending on what is actually optimized in our evolutionary model, one differentiates among three types of micro evolution

Genetic algorithms (GA)	These yield an optimization of the genotype of a species. An example is given below.
Evolution strategy (ES)	The goal of ES micro evolution is the optimization of phenotypes, say technical gadgets. A famous example is the determination of an optimal lens shape (Rechenberg, 2007).
Genetic programming (GP)	Here, the goal consists of evolving computer programs – which, for brevity, we will not discuss in the present paper (see Nash, 2007).

14.3.1 Evolution Strategy for Throwing

To demonstrate some of the features of computational micro evolution, we consider a simple task. We want to throw a ball such that it hits exactly 10 meters from our stand point, and we want to do so with the smallest energy possible. While this is a trivial task when neglecting friction, we will rather tackle the more difficult problem of throwing a "hairy" ball that has air friction $b\mathbf{v}$ proportional to velocity $\mathbf{v}$. The second order equation of motion can be rewritten as two first order differential equations for the vectors $\mathbf{x}$, $\mathbf{v}$:

$$\begin{aligned} \frac{d\mathbf{x}}{dt} &= \mathbf{v} \\ \frac{d\mathbf{v}}{dt} &= -g\mathbf{e_z} - b\mathbf{v} \end{aligned} \tag{14.11}$$

where $g = 10$ is the (downward) acceleration due to gravity and friction strength is $b = 1$. These differential equations are solved with an initial velocity (= the genotype) defined by absolute value $|\mathbf{v}_0|$ and angle α

$$\mathbf{v}_0 = (v_{01}, v_{02}) = (|\mathbf{v}_0| \cos(\alpha), |\mathbf{v}_0| \sin(\alpha)) \tag{14.12}$$

and each solution is considered an individual of the species. The phenotype, i.e. the realization of the throw then consists of the total energy used and the trajectory or throwing arc. While the energy is a straightforward function of the genotype, e.g. for unit mass

$$E = \frac{1}{2}\mathbf{v}_0^2\,, \tag{14.13}$$

this is more difficult for the throwing arc. To obtain it, the differential equations must be solved and then one has to obtain the proper section of this solution with the ground level in order to really get the endpoint of the arc. This therefore is a good example for one of the many problems of micro evolution – it is *not in general* easy to obtain the phenotype from the genotype.

The evolution algorithm for our simple model then runs as follows:

1. Start with some arbitrary velocity and angle, solve the differential equations and calculate the energy and distance from the endpoint of the arc to the target point.
2. While the iteration is running, use independent random variations of the initial throwing velocity and angle in the range of $[-1, 1]$ and perform a new solution of the differential equations.
3. If, for the new solution, the distance of the endpoint to the target point is larger than 0.02 but smaller than before, accept these variations into throwing velocity and angle as permanent.
4. If the distance to the target point is less than 0.02 and the energy of the throw is less than before, accept the variations in throwing velocity and angle as permanent.

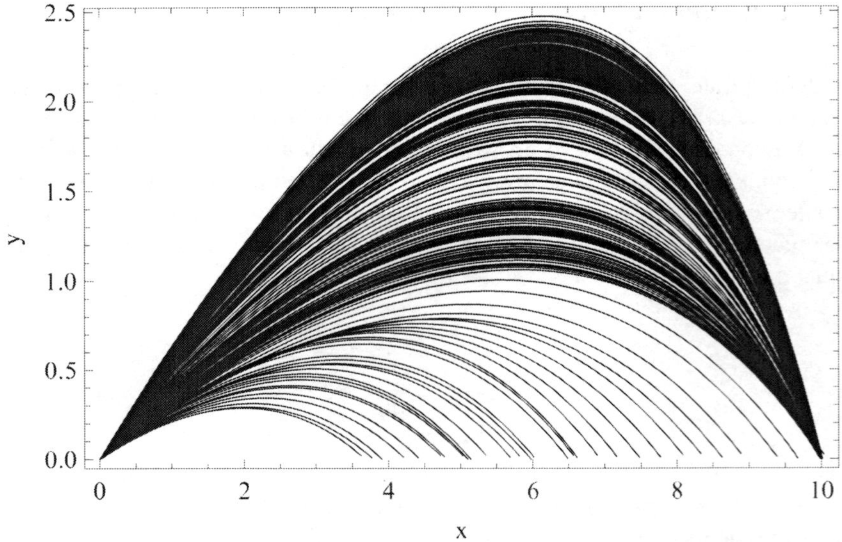

Fig. 14.4 The manifold of increasingly better throwing arcs. The system first starts to find the target point, mainly through changing the initial velocity. After acquiring target, the energy is minimized by going to a steeper angle

5. Stop the iteration either after a certain number of loops, or when a sufficient accuracy is obtained.

Figure 14.4 shows the manifold of throwing arcs. Note that from 10.000 iterations, only 344 $\approx$ 3% produced "better" results, consequently survived selection and are therefore shown in the figures – the *majority* of genetic variations had no or negative effect. Consequently, we experience a straigtforward improvement, as depicted in Fig. 14.5: After the target has been hit (i.e. distance less than 0.02, at the price of increasing v), the algorithm starts to minimize the energy and thus α and v "move" along an approximate parabola until hitting bottom. Clearly, the micro evolution thus obtained has a direction and a finite goal.

However, one has to keep in mind that the complicated space spanned by input parameters and results could, in principle, be structured in such a way that any finite number of small improvements does not allow to reach the optimal solution. The only method to get around this obstacle is to allow for nonzero survival property also in case a mutation worsens the result, i.e. to lessen the selection pressure.

If one allows, that in 30% of all cases also such neutral or negative mutations survive, one obtains the parameter history depicted in Fig. 14.6. The qualitative difference to Fig. 14.5 is striking. With relaxed selection pressure the simulated micro evolution no longer has a *direction* and a *goal*. Conversely, even after 10.000 generations – out of which more than 5.000 survive in the now friendlier environment – the second system *fills* the possible parameter space.

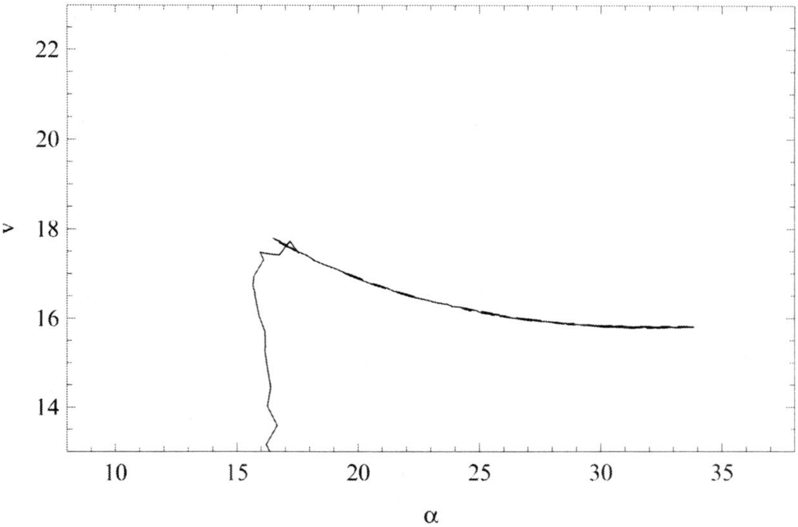

Fig. 14.5 Evolution of the parameters α (throwing angle) and v (throwing velocity) when strong selection is applied, clearly showing that artificial micro evolution in this case has a direction and a goal

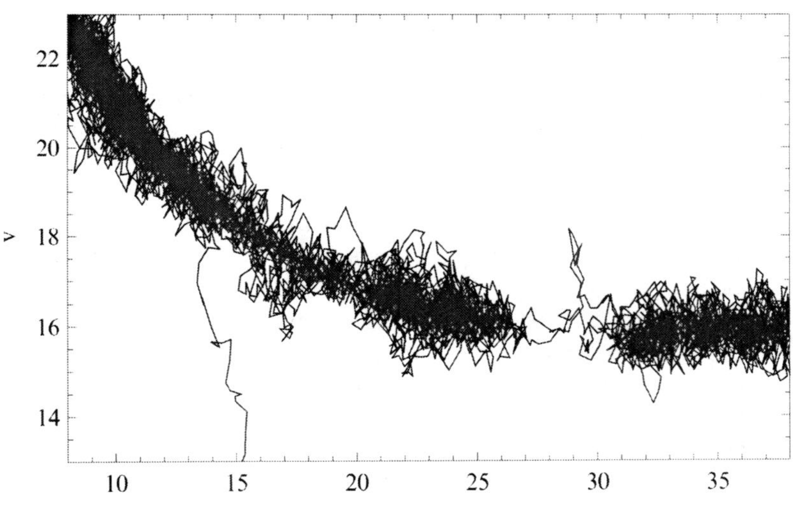

Fig. 14.6 Evolution of the parameters α (throwing angle) and v (throwing velocity) with reduced selection pressure. Direction and goal are lost, the system instead fills its parameter space and even after 10.000 generations is wandering far from an optimal solution of $\alpha \approx 32.9$ and $|\mathbf{v}_0| \approx 15.8$

While it would be difficult to show, that the second system is ergodic, i.e. comes close to any point of the parameter space, it is at least plausible that it would also find "branches" and "islands" accessible in the course of time. These two figures therefore demonstrate a very important difference between natural and artificial micro evolution:

Natural evolution — has *no* inherent direction, and lacks predictability (Gould, 2002). In fact, it is exactly this quality of filling its parameter space that enables natural ecosystems to react on critical developments. Even to "repair" the damage from catastrophic events through speciation, as we have seen in Sect. 14.2.

Micro evolution — as defined here, e.g. as the optimization of concepts by evolution strategy (random variation and strong selection) has an inherent direction (towards improval) – and is predictable in reaching a final goal.

In other words: artificial micro evolution, which is tailored to the improval of concepts (genotypes), their realization (phenotypes), strategies (behavioral patterns) or simply ideas, is completely different from natural evolution – and, strictly spoken, has nothing to do with evolution.

14.3.2 Other Examples for Micro Evolution

Once we have accepted the fundamental relationship between discrete optimization and evolution as outlined above, we can easily understand why many other models are counted as micro evolutionary models.

A simple genetic algorithm may also be used to optimize strategic decisions. Pioneering work in this field of application has been published by Axelrod (1986, 1997), and we will quote a simple example from his work. Consider the prisoner's dilemma well known in economical science: Two prisoners are pressed to testify against each other and have to decide if they yield to this pressure (decision: "defect") or refuse to do so (decision: "cooperate" - with the fellow prisoner). Payoff for this binary decision is a possible reduction of prison sentence – and this payoff may be symmetrically or asymmetrically distributed over the two prisoners. If only one round of this game is played, the strategic decision of each prisoner must be based on a guess. However, consider the case where one plays several rounds – and remembers, say, the last three games. The binary strategic decision in this case can be based on the previous results. Since these encompass $4^3 = 64$ possibilities, one has 2^{64} different strategies, or even $2^{70} \approx 10^{21}$ strategies when also recording the actual decisions made. To relate this to modern computational capabilities: 10^{21} is roughly the number of memory bits of all computers on earth in 2005.

Clearly, one cannot perform an exhaustive search for the best strategy in the iterated prisoner's dilemma. Instead, one may vary a given algorithm to achieve the best average results for a prisoner – of course, in a given environment of the other prisoner's strategy. It then turns out, that the optimal algorithm is a simple "Tit for Tat" strategy: One should offer cooperation, and punish the opponent in case he defects by also defecting.

Another example of a particularly successful research direction in micro evolution is *ant colony optimization*, a specialization of *swarm intelligence*. Its main focus lies on discrete optimization problems: Simulated "ants" are made to crawl along the links of a directed graph, and leave trails of pheromones. These are also subject to evaporation – hence after some time those pathways through the graph have the highest pheromone density that are shortest.

Ant colony optimization has been applied successfully to a large number of difficult discrete optimization problems including the travelling salesman problem, the quadratic assignment problem, scheduling, vehicle routing, etc., as well as to routing in telecommunication networks. Another interesting approach is that of particle swarm optimization, that focuses on continuous optimization problems. Here too, a number of successful applications can be found in the recent literature. Swarm robotics is another relevant field. Here, the focus is on applying swarm intelligence techniques to the control of large groups of cooperating autonomous robots.

14.4 Bottom-Up Evolution by Digital Biochemistry

We now switch to biochemistry, a completely different scale of understanding biological systems – many orders of magnitude smaller than even the level of individuals. In the past decades we have assembled a host of knowledge on the biochemistry of the cell. Apparently it contains complex and nonlinear interactions among some 10.000 relevant variables like e.g. concentrations of biochemical agents. The most formidable task of modern theoretical biology consists of modeling these interactions to the end of obtaining a "cell" that lives.

Many of the interactions among the biochemical agents are, to a certain extent, described by differential equations. On one hand this creates a big problem, since even with modern computers one cannot really solve such huge systems properly. On the other hand, the extent to which these biochemical interactions are really described by continuous differential equations seems rather doubtful. We must keep in mind, that these differential equations always refer to some continuum limit – whereas in a real cell the interactions are local and "lumpy" due to the molecular structure of matter. Of course, the same problem occurs in economical systems – where one has to deal with discrete humans instead of a continuous humanity.

On the lowest level it is therefore a problem relevant for evolutionary biology as well as for economical sciences, how to create workable models that are inherently discrete but have a continuum limit described by differential equations. Such systems are, in the language of informational science, called *state automata*. Note, that the same problem is known in theoretical physics for some time and there has been solved in terms of so called *lattice gases* that are nothing but cellular automata (see Doolen et al. (1991) for an excellent introduction). Here, we present a useful method to construct such automata out of given differential equations (Laubenbacher and Stigler, 2004). Suppose, that our biochemical system is described by a set of N coupled first order differential equations

$$\frac{dg_i(t)}{dt} = G_i(g_1,\dots g_N);\; i = 1,\dots N \tag{14.14}$$

g_i are the concentrations of the biochemical agents, G_i are nonlinear functions of the concentrations. The systems we are aiming to describe have a large N, typically of order 10.000. Nevertheless, we can understand the next steps also when considering a very simple example with $N = 2$:

$$\begin{aligned}\frac{dg_1(t)}{dt} &= 0.01\left(1 - \frac{g_1(t)}{0.01 + g_2(t)}\right) - g_1(t) \\ \frac{dg_2(t)}{dt} &= 1.5\left(1 + \frac{g_1(t)}{0.01 + g_1(t)}\right) - g_2(t).\end{aligned} \tag{14.15}$$

While it is almost trivial to solve these two differential equations (see Fig. 14.7) for a given set of initial conditions, one cannot hope to find an efficient solution for a system of magnitude $N = 10.000$. The term "efficient" hereby is an important restriction, because even if a solution of such large systems is possible, it will be consuming so many calculational resources that the formidable task to simulate the biochemistry of a single cell cannot be achieved.

We therefore replace a known solution $g_i(t)$ by a discrete approximation $\gamma_i(t)$, see Fig. 14.7. In its simplest form, this discrete approximation consists of rounding

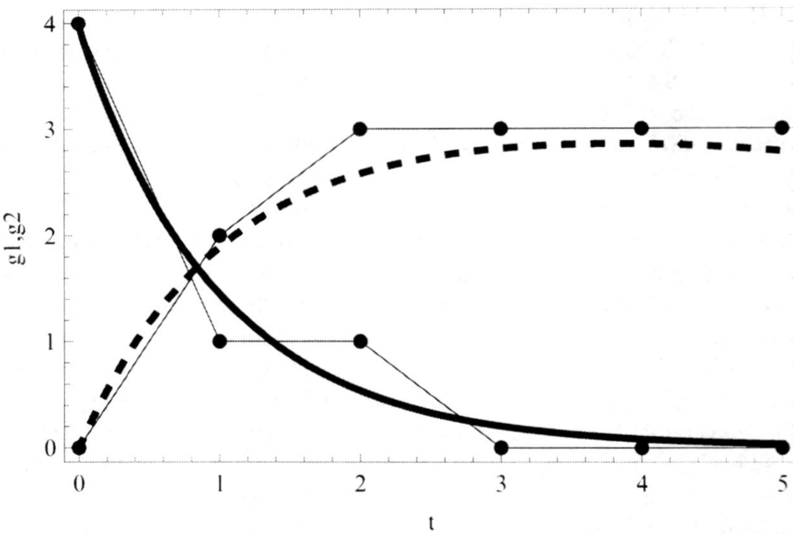

Fig. 14.7 Solution of a simple set of two coupled differential equations from (14.15) (g_1 solid line, g_2 dashed line) and the discretizations γ_1, γ_2 of this solution

to the nearest integer, but one could also use other discretization schemes. The time evolution of the two biochemical agents then may be described as a discrete time series

$$\{\gamma_1, \gamma_2\} : \{4,0\} \rightarrow \{1,2\} \rightarrow \{1,3\} \rightarrow \{0,3\} \rightarrow \{0,3\} \rightarrow \{0,3\}. \qquad (14.16)$$

The discretization now allows to implement the arithmetical structure called a *field*. Fields have acquired tremendous importance in coding theory and cryptography, for our purpose however its is sufficient to know that they exhibit a modular arithmetic. In our example the values obtained in discretization belong to the set $\{0,1,2,3,4\}$, which allows to work in the field F_5. Consequently, every calculation now is done modulo 5: $3 \times 4 = 2$, $3 + 2 = 0$ etc. are some of the obvious (albeit strange) rules of calculation, another less obvious rule is

$$\forall \gamma \in F_p : \gamma^p = \gamma, \qquad (14.17)$$

with p prime. Also not quite obvious is the fact that on a field *any* function is a polynomial of degree $\leq (p-1)$ in any of the variables. Consequently it is possible to find polynomial functions $\Gamma_i(\gamma_1, \dots \gamma_N)$ such that

$$\Gamma_i(\gamma_1(t), \dots \gamma_N(t)) = \gamma_i(t+1). \qquad (14.18)$$

This means, that there exist polynomial functions that yield exactly the time evolution of our biochemical system. Because of the particularities of the modular arithmetic, we can also be sure that these functions have integer coefficients. For the example with $N = 2$, we find one solution as

$$\begin{aligned} \Gamma_1 &= 2\gamma_1^4 + \gamma_1^2 - \gamma_2\gamma_1 + 2\gamma_2^4 - 2 \\ \Gamma_2 &= -3\gamma_2^4 + \gamma_2^2 - \gamma_1^3 + \gamma_1 + 2. \end{aligned} \qquad (14.19)$$

Figure 14.8 shows the state diagram of this automaton. The sequence from (14.16) may be found in this state diagram. It starts in the middle of the state diagram with $\{4,0\}$ and then runs into the fixed point $\{0,3\}$.

It is now very interesting to compare our discrete model to the "real" system described by (14.15). While the continuous system has a fixed point at $\{0.01, 2.25\}$ – this occurs as $\{0,3\}$ in the discrete model. The difference may be attributed to the crudeness of our approximation – we should have run it past $t = 5$. The most important aspect of this very simple example is, that it shows a multitude of initial conditions always leading to the same fixed point solution after sufficient time. Consequently, we have achieved to find a state automaton that a least for the majority of initial conditions gives a result compatible with the continuous system.

However, the discrete system also contains two further fixed points, namely $\{2,2\}$ and $\{3,1\}$ and an "oscillatory," or cyclic solution (see Fig. 14.8). These are *not* present in the continuous system – rather, any of the initial conditions for the continuous system evolves into the fixed point $\{0.01, 2.25\}$. Hence, our simple example has produced artefacts.

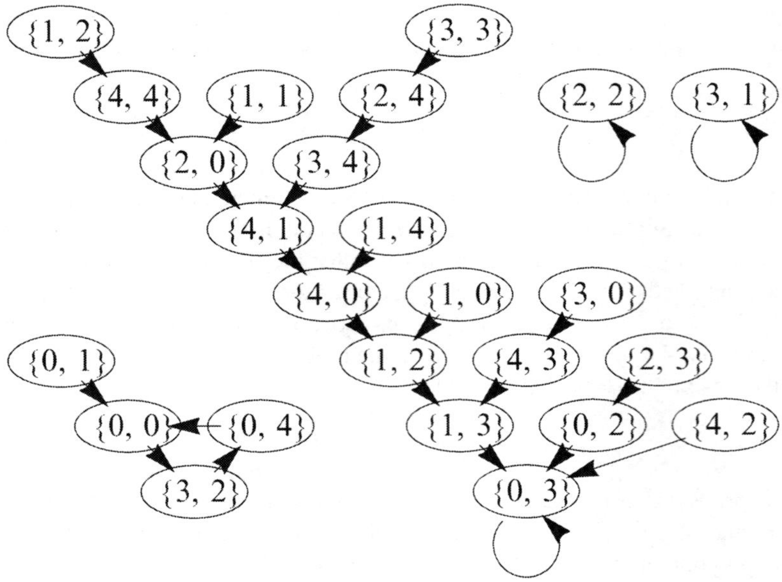

Fig. 14.8 State automaton for the differential equations from (14.15)

To remove these, we either have to include more data points, or to increase the number of discretization levels (possibly along with a rescaling of the quantities under consideration). While we do not have a systematic way to ensure a given quality of the resulting discrete model, we have at least a systematic way to generate new polynomial trial functions. This then allows to finalize the report on bottom–up evolution with the following comments.

1. Complex biochemical systems may be described, to a desired accuracy, by state automata with a finite number of states. Real biological systems, even if describing only simple regulatory circuits, generally have more than two variables – but nevertheless can be modeled using the method presented here (Julius et al., 2006; Yildrim and Mackey, 2003). Therefore, a simulation of such systems may be based on discrete variables instead of continuous ones – which automatically introduces fluctuations and a molecular "lumpiness" (Zhu et al., 2004).
2. The calculational effort to simulate systems in term of state automata is much less than in terms of continuous variables. This is due to the "bit democracy" of state automata – the corresponding computer programs do not need to involve floating point arithmetic. Consequently, this approach allows the efficient simulation of large systems with thousands of variables, and may even allow the simulation of cell arrays. Similar results are known from theoretical physics, where the usage of lattice gas models has led to simulations of dynamical processes on very large spacetime grids.
3. Since nature in itself is "lumpy" and molecular, one may argue that a discrete approach is even more appropriate to biochemistry that a continuous one. While

this is certainly true, it is also known from physics that localization and transport effects are more important than the discrete nature of matter – and therefore should be included as well.

So far, no state automaton exists for a large biological system. Hence at the moment one may only speculate about "complexity" in the resulting description. However, as we have seen in our example on macro evolution, complexity on one level does *not* require complexity on lower levels. Rather, only very simple changes in the reaction to environmental changes are sufficient to produce complexity. And as a matter of fact, such simple changes are present even in the simple system we haven chosen as an example for the present section.

14.5 Summary and Outlook

We have presented three different views on computational evolution. In considering *macro evolution*, we have produced a result with direct applicability to economical systems: Catastrophic events do not need large causes, they may be part of the dynamical behavior. Moreover it was shown, that such complex behavior does not require complex interactions among the constituents of the system. It is therefore conceivable that a similar connection model may be devised for a large economical system, say of the worldwide stock market. Since such a system would include the same main ingredients as the one presented in Sect. 14.2, i.e. random fluctuations and collective behavioral change, we may safely expect that it also exhibits catastrophic events without external cause (Hand and Henning, 2004). Conclusions are left to the reader.

In our Sect. 14.3 on *micro evolution*, we have relied on existing standard models – genetic algorithms, evolution strategy and genetic programming. While these are useful in a multitude of situations – ranging from algorithms to improve mechanical devices, we have shown a clear distinction between natural evolution and artificial micro evolution: The latter has a direction and end a predictable goal. We found this to be the influence of the *perfection* of the selection. A perfect selection pressure leads to improvement in a given situation. However, it is well known that the evolutionary perfection also may lead to strange animals – like little blind fish, believing their little cave to be the whole world. Optimization therefore may be a disadvantage, because it disables the variety needed to accommodate to changing situations.

Natural evolution in contrast is free of direction and has unpredictable results. An interesting question in this aspect is, how much imperfection has to be built in the selection criteria of a given ecological (or economical?) system to retain a required level of adaptability. Lessening the selection pressure implies to use less information about the system and its future – not in the form of bits and bytes, i.e. syntactic information, but rather in the form of semantic information. The distinction between these two aspects of information is crucial when relating information with entropy and measurability (Henning, 2004). Since it has been shown already, that missing information has a profound influence on the quality of *optimal strategies* in economical games (Schredelseker), it might be worthwhile to study how it affects the parameter space in evolution strategy for such games.

Our final Sect. 14.4 then focused on biochemistry to demonstrate, how one may simulate large arrangements of regulatory circuits. Of course, the current goal of this bottom-up evolution is the simulation of living cells (and possibly their interaction). However – and here again we quote the multi-scale view on evolution by Stephen Jay Gould – the same methods may be used on *any* level. One could devise finite state automata also for macro evolution, say, to replace the Lotke–Volterra differential equations governing predator-prey relations. It is conceivable, that one may also simulate regulatory circuits of economical systems with such a model.

The author wishes to end the paper with a prediction: As lattice gas models have found their niche in theoretical physics, cellular and state automata are firmly established in evolutionary biology and beyond, they will find increasing usage also in economical and social sciences. Some steps into this direction have been made already – like e.g. the modeling of traffic behaviour, or human panic behavior by cellular automata.

References

R. Axelrod. An evolutionary approach to norms. *American Political Science Review*, 80:1095–1111, 1986.

R. Axelrod. *The Complexity of Cooperation*. Princeton University Press, Princeton, NJ, 1997.

J. M. Baker. Adaptive speciation: The role of natural selection in mechanisms of geographic and non-geographic speciation. *Studies in History and Philosophy of Biological and Biomedical Sciences*, 36:303–326, 2005.

R. K. Bambach, A. H. Knoll, and S. C. Wang. Origination, extinction, and mass depletions of marine diversity. *Paleobiology*, 30(4):522–542, 2004.

G. D. Doolen, U. Frisch, and S. Wolfram, editors. *Lattice Gas Methods for Partial Differential Equations*. Addison-Wesley, Reading, MA, 1991.

B. Eichengreen and H. James. *Gold Standard and the Great Depression, 1919-39*. Oxford University Press, Oxford, 1996.

N. Eldredge and S. J. Gould. Punctuated equilibria: an alternative to phyletic gradualism. In T.J.M. Schopf, editor, *Models in paleobiology*, pages 82–115. Freeman, Cooper, San Francisco, 1972.

G. B. Ermentrout and L. Edelstein-Keshet. Cellular automata approach to biological modeling. *Journal of Theoretical Biology*, 160:97–133, 1993.

J. K. Galbraith. *The Great Crash: 1929*. Houghton Mifflin, Boston, 1997.

S. J. Gould. *The structure of evolutionary theory*. Belknap Press of Harvard University Press, Cambridge, 2002.

I. Hand and P. A. Henning. Glücksspielen an der Börse. eine verhaltenspsychologisch-mathematische Analyse. *Sucht*, 50(3):172–186, 2004.

P. A. Henning. Zum Informationsbegriff der Physik. *Informatik Spektrum*, 27(2):202–207, 2004.

A. A. Julius, A. Halasz, V. Kumar, and G. J. Pappas. Finite state abstraction of a stochastic model of the lactose regulation system of escherichia coli. In *Proceedings 45th IEEE Conference on Decision and Control*, 2006.

R. Laubenbacher and B. Stigler. A computational algebra approach to the reverse engineering of gene regulatory networks. *Journal of Theoretical Biology*, 229:523–537, 2004.

R. A. Muller and R. A. Rohde. Cycles in fossil diversity. *Nature*, 434:208–210, 2005.

J. R. Nash. Simulated evolution. an open source project. Cited April 2008, 2007. URL `http://www.simulatedevolution.com`.

I. Rechenberg. Evolution strategy to find an optimal lens shape. Cited April 2008, 2007. URL `http://lautaro.fb10.tu-berlin.de/user/michael/english/lens/lens.html`.

W. R. Rice and E. E. Hostert. Laboratory experiments on speciation: What have we learned in forty years? *Evolution*, 47:1637–1653, 1993.

K. Schredelseker. On the value of information in financial decisions – a simulation approach. University of Innsbruck.

R. Solé and S. Manrubia. Extinction and self-oranized criticality in a model of large-scale evolution. *Physical Review*, B54(8):R42–R45, 1996.

N. Yildrim and M. C. Mackey. Feedback regulation in the lactose operon: A mathematical modeling study and comparison with experimental data. *Biophysical Journal*, 84:2841–2851, 2003.

H. Zhu, P. Pang, Y. Sun, and B. Dhar. Asynchronous adaptive time step in quantitative cellular automata modeling. *BMC Bioinformatics*, 5:85, 2004.

Chapter 15
Artificial Markets: Rationality and Organisation

Alan Kirman

"*It is a peculiar fact that the literature on economics contains so little discussion of the central institution that underlies neoclassical economicsthe market.*" (North, 1977, p. 710)

Abstract Economics models markets as mechanisms linking anonymous and otherwise isolated individuals through the price system. Little attention is paid to how these markets are organised nor as to how agents interact within them. In particular it is interesting to know how agents following rather simple rules come to coordinate. This paper discusses the modelling of these important aspects of markets and suggests that building artificial markets can complement the usual theoretical, empirical or experimental approaches. Using specific examples of empirical markets the relationships that develop between agents can, for example, be modelled and the results from the simulated markets compared with the stylised facts. Artificial market models are neither ad hoc nor theory free, but thys allow us to analyse situations which are too general to permit analytic solutions.

15.1 Introduction

Economic theory is wedded to two basic ideas. Firstly there is the idea that the rationality of economic agents obeys the axiomatic structure typified by that used in the Arrow Debreu model. Secondly, it is assumed that market organisation as such has little impact on economic outcomes. Individuals participate in anonymous markets in which they are price takers and little is said about who sets the prices and how. When exceptions are made to this basic structure it is significant that economists refer to "imperfect competition" and market "imperfections." Thus there

A. Kirman
GREQAM, Université de la Méditerranée, 2 rue de la Charité, 13002 Marseille, France, e-mail: alan.kirman@univmed.fr

is a benchmark model and other situations in which individuals react to each other are thought of as deviations from the norm. In the most basic theory, the General Equilibrium model, which traces its origins to Walras and Pareto, individuals interact only through the price system. Direct interaction, and the externalities that go with it are either declared to be the subject of game theory or are incorporated with difficulty into a modified GE model.

In this brief paper I would like to argue that we should turn things inside out and bring direct interaction to the center of the stage. Furthermore, I claim that we should radically simplify our models of individuals and that in so doing we may still observe interesting and complicated aggregate behaviour which is, however, the result of the aggregation itself and not of the complicated behaviour of some "representative individual." We should treat the economy as a complex system and as in other disciplines we should not expect the system to behave like an individual.

The way to do this is by building models of simple individuals who follow simple rules and interact with each other just as molecules in biological systems or particles in physical systems. The usual argument against this is that humans have intentions and are forward looking and cannot therefore be modeled as one would model molecules or inanimate particles. This misses the essential point, if we can describe the rules that an individual follows and the way, in which he interacts with other individuals, we can use our models to understand what the outcomes will be. We do not need to know what the deep psychological motives for an individuals actions are. Consider the argument that individuals are forward looking, and think of the problem of forecasting. In all our models individuals map past history into forecasts of the future. Once they have a forecast of the future they take an action, so in the end, they just have an algorithm, which maps past history into actions. There is nothing intrinsic which prevents us from building simple agents or robots that do this. We can choose what we consider to be the appropriate level of sophistication for the mapping from the past to actions; we can also model the reactions of other agents to an individuals choice of actions. What is more we can let the agent learn about the rules that he uses and we can find out if our simple creatures can learn to be the sophisticated optimisers of economic theory.

In doing this we are not straying far from what has been recommended by our illustrious predecessors and the leaders of our profession. The first idea that I am suggesting is that we have to treat the economy as a complex system. But Herb Simon (1962) already described a complex system when explaining how the thought economic theory should develop and he said:

"Roughly by a complex system I mean one made up of a large number of parts that interact in a nonsimple way. In such systems, the whole is more than the sum of the parts, not in an ultimate metaphysical sense, but in the important pragmatic sense that, given the properties of the parts and the laws of their interaction, it is not a trivial matter to infer the properties of the whole. In the face of complexity, an in-principle reductionist may be at the same time a pragmatic holist." Herbert Simon (1962, p. 267).

The second argument that I make is that we should dispense with the a priori assumptions about rationality and optimisation, which are so central to economic theory. But, if you think that this might be heretical consider what Robert Lucas (1988), had to say on the subject:

"In general we view or model and individual as a collection of decision rules (rules that dictate the action to be taken in given situations) and a set of preferences used to evaluate the outcomes arising from particular situation-action combinations. These decision rules are continuously under review and revision: new decisions are tried and tested against experience, and rules that produce desirable outcomes supplant those that do not. I use the term "adaptive" to refer to this trial-and-error process through which our modes of behaviour are determined."

However, Lucas then goes on to argue that we can safely ignore the dynamics of this process since,

"Technically, I think of economics as studying decision rules that are steady states of some adaptive process, decision rules that are found to work over a range of situations and hence are no longer revised appreciably as more experience accumulates."

In general, however, one cannot assume convergence to some equilibrium but one has to look at the dynamic evolution of the economy resulting from the interaction between agents. One is also interested in knowing how the state of the system evolves over time and not only whether it settles down to what might be thought of as some sort of equilibrium. Here I am taking a different position from Lucas and arguing that one cannot assume that all the adaptation has taken place in the past but that we are faced, in economics, with many situations in which individuals are constantly adapting to change and thereby generating change. Thus, not only the relevant time scale but also the process itself is very different from that relevant for biological evolution, which is too often used by simple analogy.

15.2 Relationships in Markets

"Applications of economic theory to market or group behaviour require assumptions about the mode of interaction among agents as well as about individual behaviour" Lucas (1988).

Think of the standard vision of a market, which is a system in which the actors act according to a system of rules, which constrains them, and that this generates the aggregate economic outcomes. They are anonymous and their relations with others are not considered. Those who participate in, who regulate or study actual market mechanisms have a very different view. For example Aboulafia argues that markets are essentially social institutions in his well-known study of financial markets, indeed he says,

"Markets are socially constructed institutions in which the behavior of traders is suspended in a web of customs, norms, and structures of control...Traders.negotiate the perpetual tension between short-term self-interest and long-term self-restraint that marks their respective communities" Aboulafia (1997).

Kuhn goes further and argues that individual relationships and trust are necessary for the functioning of markets. For him, it is clear that,

"Markets are not self-operating, objective mechanical objects. They are, rather, a complex set of constraints, rules, rights, regulations, and laws, guiding human participants in making their multiple, various trades, purchases, and exchanges. The motivating force that generates benign market outcomes is the willingness of all to obey the guidelines and deal openly transparentlywith each other. Invisible to the naked eye are the common social bonds of trust among all, strangers and acquaintances alike. The bonds of trust are what create and sustain truly efficient, effective markets." Kuhn (2005).

In another context Alan Greenspan, Chairman at the time of the Federal Reserve, has remarked that,

"It is hard to overstate the importance of reputation in a market economy. To be sure, a market economy requires a structure of formal rules–a law of contracts, bankruptcy statutes, a code of shareholder rights–to name but a few. But rules cannot substitute for character. In virtually all transactions, whether with customers or with colleagues, we rely on the word of those with whom we do business. If we could not do so, goods and services could not be exchanged efficiently. Even when followed to the letter, rules guide only a small number of the day-to-day decisions required of corporate management. The rest are governed by whatever personal code of values corporate managers bring to the table." Greenspan (2003).

This poses a problem for those who would like to model the way markets really function. Whilst the anonymous market poses few problems, one just has to specify the rules which individuals follow when they are faced with the prices given by some unspecified market mechanism, but here we are faced with the idea that individuals build up relations of confidence with each other and this seems more like a subject for psychologists. If we specify who interacts with whom we simply have to specify the graph in which the various buyers and sellers are linked. However, if we are to explain how these links form and are sustained the task is more difficult. Nevertheless, I would like to argue that we can create models in which individuals learn with whom they wish to interact. For this reason, I will focus in the first part of the paper on the development of relationships in markets and, in particular in a market, the wholesale fish market in Marseille for which we have very complete data. In particular I will argue that artificial markets can contribute to the quest for the explanation of this and some of the other features of the complex market structures that we observe.

However, artificial markets on their own are what their name suggests, artificial. The three traditional approaches for economists are, theoretical, experimental and empirical. The idea here is that the fourth approach that I shall analyse, that of using agent based models to construct artificial markets which are then simulated, can complement the other approaches each of which has its weaknesses.

What are the drawbacks of theoretical models? The first and most important weakness is that they have to be greatly simplified in order to make them analytically tractable. That is, we have to reduce them to a minimum to be able to solve them analytically, for example to characterise the equilibria. The second is that the assumptions are often made for analytic tractability rather than for economic realism. Artificial markets can help here by providing results in more general analytically intractable situations and then seeing if these results obtained in the simpler case.

What are the limitations of experiments? Once again, in order to make the situation understandable for the subjects one has to simplify. Furthermore, the situation with which the subjects are faced is extremely unnatural. Often they believe that they have a problem to solve for which there is a right answer, thus rather than reacting naturally they try to outguess the experimenter. Thus one would like to know if, other than using simplified theoretical models, one could not develop another benchmark against which to evaluate subjects behaviour in experiments. Again artificial models can be used to provide such a benchmark.

Finally, why can one not content oneself with working with the data from markets directly. Doing this can enable us to establish some "stylised facts" or statistical regularities but gives us no idea as to the structure that generated them. To get an understanding of how the system functions we typically build a model and then we are faced again with the choice between a theoretical model and its agent-based counterpart. The suggestion here is that we can usefully employ both approaches. What I will suggest is that we should use empirical markets as our benchmark, and then construct models, which reproduce some of the salient features of these markets. I will start by looking at a particular market, the wholesale fish market in Marseille for which we have very detailed data.

15.3 The Marseille Fish Market (Saumaty)

At the time that the data was collected the wholesale fish market for Marseille situated at Saumaty on the coast at the Northern edge of Marseille, was open every day of the year from 2 to 6 a.m.[1], over 500 buyers[2] and 45 sellers came together, although they were not all present every day and they transacted more than 130 types of fish. Prices were not posted. All transactions were pair wise. There was little negotiation and prices can reasonably be regarded as take it or leave it prices given by the seller. The data set consists of the details of every individual transaction made over a period of three years. The data was systematically collected and

[1] Things have changed now. Unfortunately for Annick Vignes who was responsible for obtaining the data, and myself, since I went there from time to time and she visited it regularly at the times mentioned, the market is now open during the day rather than at night.

[2] There are 1,400 buyers in the records but many of these were hardly present at all.

recorded by the Chambre de Commerce de Marseille, which managed the market at that time. The following information is provided for each transaction:

- The name of the buyer
- The name of the seller
- The type of fish
- The weight of the lot
- The price per kilo at which it was sold
- The order of the transaction in the daily sales of the seller.

The data is for the period from the 1st of January 1988 to the 30th of June 1991. The total number of transactions for which we have data is 237,162.

Although no auction mechanism has been set up, when the market was being reorganised, provision was made for the establishment of an auction. Neither buyers nor sellers were favourable to this development and so the market has remained organised as previously. There are two possible reasons that one could invoke for rejecting the change to an auction. Firstly, the buyers and sellers may simply have thought that an auction would be less efficient. Secondly and more plausible, they may just have been afraid of operating in a system with which they were not familiar. This seems the most obvious explanation for the sort of inertia that is often observed in institutional arrangements. Finally, the role of the actors and the actors themselves might change with the new mechanism. The sellers currently offer 90% of their fish from sources other than the Mediterranean Sea. This fish is bought at auction from other ports. Would these sellers be able to buy like this and then to put the fish up for auction a second time or would shortcuts develop between the major buyers and the other markets?

Whatever the reasons, the current arrangement in Marseille gathers together a large number of buyers and sellers who negotiate with each other over a stock of fish, which is already determined at the opening of the market. Given the number of agents involved in the same place at the same time on a regular basis one might be led to expect that, after a while, prices would converge, in the sense that the same type of fish would be sold at essentially the same price on a given day. This, as I will show, is far from being the case. Furthermore, one might expect to see evidence of the standard assertion that, as the day progresses, prices diminish. This is again not true. Given that all the buyers and sellers are gathered in one place it would seem to be more or less costless for buyers to check the offers of different sellers before making their purchases. Yet what one observes is that a large proportion of buyers frequent only one seller. This reflects the argument made previously that relations build up between buyers and sellers, perhaps based on trust. Now the question is how one might model the development of these relationships. Trust may, in fact, be a synonym for previous successful experience and with this in mind I will sketch as simple model, originally developed with Gerard Weisbuch and Dorothea Herreiner, (Weisbuch et al., 2000). This model shows how specific relationships can emerge as a result of reinforcement learning. The basic, extremely rudimentary, theoretical model is then simulated in more general form to see whether the stylised facts generated in the simplest form hold up. These facts can then be compared with the empirical evidence. After this I shall discuss models with less sophisticated learning to see whether they can generate other important features of the real market. Then I shall outline a model for a market that functions on a different basis, using an auction rather than pair wise trading. Here we have detailed data from the Ancona fish market, which operates on the basis of three simultaneous Dutch auctions.

15.4 A Simple Market Model

In accordance with economic theory one might be tempted to try to find the equilibrium in a fully optimising game-theoretic model of the market. However, whilst such a model may allow one to show that equilibrium exists, it is very difficult to characterise such equilibrium. Furthermore the sophisticated behaviour of the agents in such a model is a poor description of reality. For these reasons, we have to turn to a different type of model. Assume that agents use simple rules to make their choices. If this is the case then the notion of "equilibrium" outcome has to be redefined since it will depend on the rules chosen. It may be the case however, that this sort of approach based on more rudimentary behaviour may be more effective in reproducing the sort of phenomena we observe on markets. In particular there are a number of phenomena which are of interest in their own right and which can be examined in this way. For example we would like to be able to analyse the sort of trading relationships that emerge. We would also like to have a model, which generates a non-degenerate distribution of prices.

As far as the first problem is concerned, let me remark that the sort of trading relationships that we observe in markets like Marseille have rather specific patterns. These must, at least in part, be responsible for the sort of price dispersion that is observed. In addition, although these networks may play an important role in determining market outcomes, nothing is said in the sort of game theoretical model discussed above, about their impact or the evolution. They could be included in that sort of model in two ways. Firstly, the network could be taken as given and this might be thought of as restricting the information available to the individuals. Secondly, and much more ambitiously, they could be formed strategically, but both of these approaches would merely make the analysis less tractable. What we need is a way of looking at the emergence of these networks, which does not make unreasonable demands on the calculating capacity of the individuals involved. Before doing this it is perhaps worth asking what we actually observe.

What is the nature of the trading relationships on the Marseille fish market? There is a remarkable dichotomy. On the one hand, there are those buyers who regularly buy from the same seller and are extremely loyal, and on the other hand, there are people who shift between sellers all of the time. Yet, there are few buyers who are basically loyal but occasionally shop around. This, itself, seems to be a feature that one should try to explain. If one tries to go back to a full game theoretic model this becomes extremely complicated because one has to develop now a dynamic game in which the experience of playing with each seller is taken into account. Alternatively, one has to think of a situation in which people have strategies, which are so complicated that they can take into account all the possible prices they might face from each of the different sellers at each point in time.

What I would like to suggest here is the idea of developing a much simpler theoretical model in which people simply learn from their previous experience and they in consequence change their probability of visiting different sellers as a result of their experience. What I will argue is that models of this sort which attribute very little computational ability or general reasoning capacity to individuals may be capable of generating specific features of real markets. This sort of "bounded rationality" approach has received a lot of attention but is often dismissed for its lack of rigour. In reality, the analysis of the evolution of the "state" of the market in the model can be perfectly rigorous given the specific choice of rules for the agents.

It is still the case that building artificial markets, in which agents have simple rules of behaviour, is not widely accepted in economics and one might wonder why. The reason seems to me simple; choosing rules of thumb for agents is regarded as ad hoc. However, we have come to accept that the restrictions that we impose on the preferences of individuals, unlike other behavioural rules, are not ad hoc. Therefore, if we replace those assumptions, which by their very nature cannot be empirically tested, by other rules, we are subject to the criticism that we lose the rigour of "proper micro foundations". Artificial markets, it is said are not "scientific." Let me simply suggest that maximisation of a well defined preference order is not necessarily a reasonable assumption when both the order to be maximised and the set of alternatives are highly complicated, and that something is to be gained from simplifying our account of individuals' behaviour in complicated situations.[3]

One response to this is that in the process of using simple rules individuals converge to those rules, which work best and therefore act just as if they were maximising in the standard way. This is Lucas (1988) position. There are two objections to this. First, we have to show that the learning process converges, and if it does that it corresponds to the maximisation in the original problem. Second, learning processes usually involve learning about something, which is not changing. But here, the learning is influenced by the behaviour of other individuals who are also learning. It is by no means clear that we will have convergence in such a situation. The answer to this is that we are interested in modelling the results of interactions between individuals following simple rules, not just as a way of justifying a theoretical equilibrium but rather as a vehicle for understanding empirical reality. Artificial markets should be a step closer to reality than their theoretical counterparts.

15.5 Trading Relationships Within the Market

As an example, consider a model, which we developed as a simplified version of, the Marseille fish market (see Weisbuch et al., 2000). There we consider a situation in which buyers do not anticipate the value of choosing sellers but rather develop relationships with sellers on the basis of their previous experience.

[3] I have developed this sort of argument at length in Kirman (2006) where I suggest that we have gone down the wrong route in modelling demand

We will consider a market in which the buyers choose their sellers with certain probabilities. Buyers do not necessarily choose, a priori, a particular seller to visit but rather increase their probability of visiting sellers who gave them good results in the past. To be more precise, there are n buyers indexed by i and m sellers indexed by j. Now, denote by $J_{ij}(t)$ the cumulated profit, that buyer i has obtained from trading with seller j up to time t, and this is obtained as follows:

$$J_{ij}(t) = (1-\gamma)J_{ij}(t-1) + \pi_{ij}(t). \tag{15.1}$$

In other words, at each point the buyer i discounts the cumulated profit he has made from seller j and adds the profit he make at that time. Now we have to specify the mapping from these past profits to probabilities. We shall assume that the buyers update their probability of visiting sellers on the basis of the profit that they obtained in the past from them as follows. The probability $p_{ij}(t)$ that i will visit j in period t is given by,

$$p_{ij}(t) = \frac{e^{\beta J_{ij}(t)}}{\sum_k e^{\beta J_{ik}(t)}} \tag{15.2}$$

where β is a reinforcement parameter which describes how sensitive the individual is to past profits. This non-linear updating rule will be familiar from many different disciplines and is also widely used in statistical physics. It is known as the "logit" rule or, in game theory as the "quantal response" rule. To reiterate, the rule is based on two simple principles. Agents make probabilistic choices between actions. Actions that have generated better outcomes in the past are more likely to be used in the future. Such a process has long been adopted and modelled by psychologists (see e.g. Bush and Mosteller, 1955). It is a special form of reinforcement learning. It has also been widely used in evolutionary and experimental game theory (see Roth and Erev, 1995) and a more elaborate model has been constructed by Camerer and Ho (1999). The particular form chosen here is used extensively in economic models with learning. It is found in the model developed by Blume (1993), for example, to analyse the evolution of the use of strategies in games. This approach has a number of advantages. In particular, it requires no specific attribution of rationality to the agents other than that they are more likely to do what has proved to be successful in the past. Furthermore, it has the property that agents always experiment with sellers that they have not tried in the past even though the probability of doing so decreases if they find suitable partners.

15.6 A Little Formal Analysis

To simplify matters at the outset we will start with a continuous approximation of our model which is actually in discrete time. Furthermore, we will replace the random variables by their expected values. This is referred to as the "mean field" approach. In this way it is easy to see that the change in cumulated profit for the buyer is given by,

$$\frac{dJ_{ij}}{dt} = -\gamma J_{ij} + E(\Pi_{ij}). \tag{15.3}$$

Using the learning rule that we have given we know the probability for agent i to visit seller j and can therefore calculate the expected gain from that visit. Recall that there are two things involved here, firstly the probability that the seller j still has fish available when buyer i arrives, and secondly the probability that the latter chooses seller j. So the expectation is given by,

$$E(\Pi_{ij}) = \Pr(q_j > 0)\Pi_{ij}\frac{\exp(\beta J_{ij})}{\sum_k \exp(\beta J_{ik})}. \tag{15.4}$$

Now consider an even simpler case where the seller is sure to have fish, in which case we have,

$$\Pr(q_j > 0) = 1. \tag{15.5}$$

Now simplify even further and look at the case where there are just two sellers and furthermore each time a buyer visits one of the sellers he receives a fixed profit of Π and find the equilibrium level for the cumulated profit for a buyer from seller 1 and this will of course be when

$$\frac{dJ_1}{dt} = 0. \tag{15.6}$$

Substituting this gives,

$$\gamma J_1 = \Pi \frac{\exp(\beta J_1)}{\exp(\beta J_1) + \exp(\beta J_2)}. \tag{15.7}$$

Now take the difference between the profits from the two sellers and we have,

$$\Delta = J_1 - J_2. \tag{15.8}$$

If we now substitute we have the following expression,

$$\Delta = \frac{\exp(\beta\Delta - 1)\Pi}{\exp(\beta\Delta + 1)\gamma}. \tag{15.9}$$

We now have simply to solve this equation for Δ and this gives two cases. First, consider

$$\beta < \beta_c = \frac{2\gamma}{\Pi}. \tag{15.10}$$

In this case, when the importance attached to previous experience is below the critical value β_c we have,

$$\Delta = 0 \quad J_1 = J_2 = \frac{\Pi}{2\gamma}. \tag{15.11}$$

There is a single solution and the cumulated profits from both sellers and hence the probabilities of visiting them are the same.

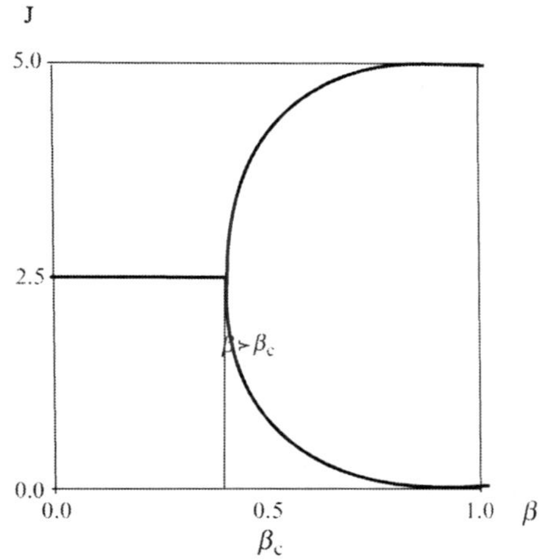

Fig. 15.1 The transition from random buying to loyalty as a function of β, (source Weisbuch et al. (2000)).

J represents the equilibrium cumulated profit from seller 1 which translates into the probability of visiting that seller, as β passes the critical value the equilibrium probability changes rapidly either becoming close to 1 or to 0

However, when $\beta > \beta_c$ then there are three solutions and $\Delta = 0$ is unstable and there is a rapid transition at $\beta = \beta_c$. By which we mean that as soon as β passes above the critical value the probabilities of visiting each seller become rapidly very different. All of this is illustrated in Fig. 15.1.

To repeat, what we can see is that as the value β that is, the importance that the buyer attaches to previous experience increases, there are two stable solutions for the probability (the relation between J and the probability is given by the equilibrium condition), that he has of visiting seller 1. When he has that probability he has the complementary probability of visiting seller 2. Thus he will spend most of his time with one seller and much less with the other. As β increases towards 1 the buyer becomes completely attached to one or the other seller. So in this very simple case we see that loyalty is a property of the equilibrium when β is high enough.

However, remember how we arrived at this solution. We built a simple approximation by using the "mean field" approach and then with a number of extreme assumptions derived the solution. The mean field approach involves making two approximations. We considered that the process was a continuous one and furthermore that we could replace the random variables by their expected values. In addition we considered the case with just two sellers each of whom gave equal profits and always had fish to sell.

It is here that artificial markets come into play. What we want to analyse is a situation in which buyers are different for example and have either different discount factors or obtain different profits. It is too difficult to find an analytical solution for the full stochastic process with several sellers and buyers with varying characteristics, but we can simulate it. Thus we create an artificial market with agents who learn as we suggested and see whether the solutions of our analysis of the simple case correspond to those of the simulations.

Perhaps, even more importantly we do not have to solve for the equilibrium but can actually simulate the learning process and see whether it converges to the theoretical equilibrium.

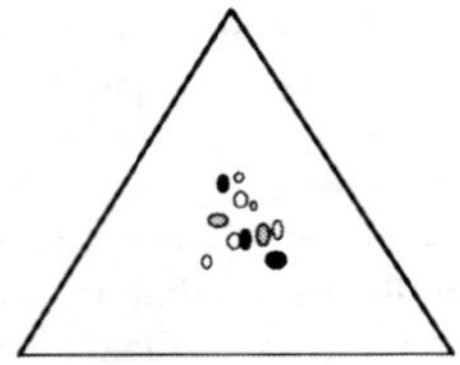

Fig. 15.2 The probabilities of buyers visiting the three sellers when β is low β, (source Weisbuch et al. (2000)).

Each dot represents the probabilities of one buyer buying from the three sellers, when the dots are in the middle of the simplex buyers are visiting all sellers with similar probabilities

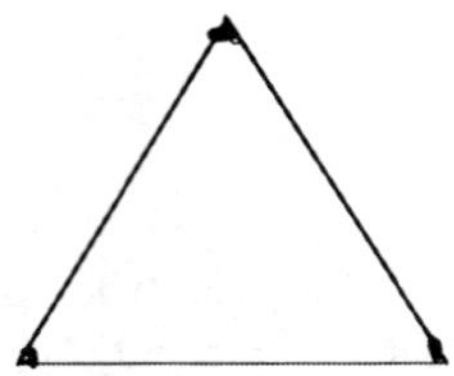

Fig. 15.3 The probabilities of buyers visiting the three sellers when β is high, (source Weisbuch et al. (2000)).

Each dot represents the probabilities of one buyer buying from the three sellers, when the dots are at an apex of the simplex buyers are visiting one seller with probability 1

Consider, as an example the case in which there are three sellers and thirty buyers in the market. At any point in time, each buyer will have a probability of visiting each of the sellers. Thus he will be represented by a point in the three simplex or triangle as illustrated in Fig. 15.2 below. Each buyer is represented by such a point in the simplex and the nature of the relationships will be illustrated by a cloud of points. A buyer who shops around in a purely random way, that is who is equally likely to visit each of the three sellers will be represented as a point in the center of the triangle. If, on the other hand, he visits one of the sellers with probability one then he can be shown as a point at one of the apexes of the triangle.

Thus, at any one point in time, the market is described by a cloud of points in the triangle and the question is how will this cloud evolve as the buyers learn? If buyers all become loyal to particular sellers then the result will be that all the points, corresponding to the buyers will be at the apexes of the triangle as in Fig. 15.3. This might be thought of as a situation in which the market is "ordered." The network of buyers and sellers becomes deterministic. On the other hand, if buyers learn to search randomly amongst the sellers, then the result will be a cluster of points at the center of the triangle, as in Fig. 15.2. What we showed in Weisbuch et al. (2000), is that which of these situations will develop, depends crucially on three parameters β, the reinforcement factor which represents the importance attached to previous experience, γ the discount rate and π the profit per transaction. The stronger the reinforcement, the slower the individual forgets and the higher the profit obtained from sellers, the more likely is it that loyalty will emerge.

It might be asked whether or not the features of the actual market in Marseille do actually reflect the sort of behaviour predicted by this, admittedly primitive model. What the model suggests is that the transition from disorder to order, or more prosaically from shopping around to loyalty as β changes, is very sharp. The change will depend on a critical value β_{ci} of β which will be different for each buyer i and will depend on the frequency of his visits to the market and his profit. It is easy to see why higher profits obtained will make a buyer stick to the seller that gave him those profits but the importance of the frequency of visits needs a little explanation. If a buyer comes infrequently to the market then his information from previous visits is less pertinent than which the regular visitor got from his last visits. He will therefore discount previous experience more than his loyal counterpart. This will lead him to reinforce his tendency to shop around. To sum up then the critical value is determined by the ratio $\frac{2\gamma_i}{\Pi_{ij}}$ and the discount factor and the profit obtained from each seller depend on the buyer in question. In our theoretical model one can vary the profits that buyers obtain from the seller and for example one can assume that one of the sellers is more profitable for the buyers than the others. In this case, contrary to what one might expect, the equilibria will give a higher probability to going to the better seller but will not lead everyone to buy from him.

Before leaving the theoretical model, one observation is in order. The patterns illustrated in Fig. 15.2 were obtained with buyers all of whom had the same β_{ci} and the same is true for Fig. 15.3 where the value of β_{ci} is, of course, lower. Recall again, however, that the critical value β_{ci} varies with the buyers. Now, what happens in the simulations if the group is mixed as it is in reality? In the loyal buyers situation, the sellers learn to sell the correct amount because they have fixed and regular customers. In the random shopper situation, sellers cannot predict accurately the amount that they will sell. Therefore there is more waste, some buyers are unsatisfied and some sellers are left with fish on their hands. Now, will it not be the case that the presence of random shoppers will interfere with the profits of the loyal and thus weaken their loyalty? Sellers will sometimes have insufficient stocks because they have been visited by random shoppers and this could have a negative effect on those who were normally loyal to those sellers. We simulated the situation with equal proportions of loyal or low β_{ci} and high β_{ci} buyers. Interestingly, the presence of the random shoppers did not prevent the development of loyalty as is shown in Fig. 15.4. Those whose low β_{ci} led them to be loyal remain loyal and those whose high β_{ci} led them to shop around continue to do so.

What then should we expect to observe on the real market if what our model suggests is right? As I have said, the important conclusion is that the division between loyal buyers and random shoppers should be quite sharp and one should not expect to find individuals who shop around to some extent but are somewhat more loyal to some sellers than to others. This is precisely what is observed on the Marseille fish market. The behaviour of buyers is highly bimodal. Consider the case of cod as an example. In Fig. 15.5 the histogram of the number of buyers visiting different numbers of sellers is shown. There is a concentration of buyers who only visit one seller and then a distribution of individuals who visit several sellers with a median of 4 per month.

As it happens it is often the case that buyers will purchase more of one fish than of another. We might then ask whether loyalty is the same across all types of fish. The extent of the loyalty of customers for certain types of fish can be illustrated

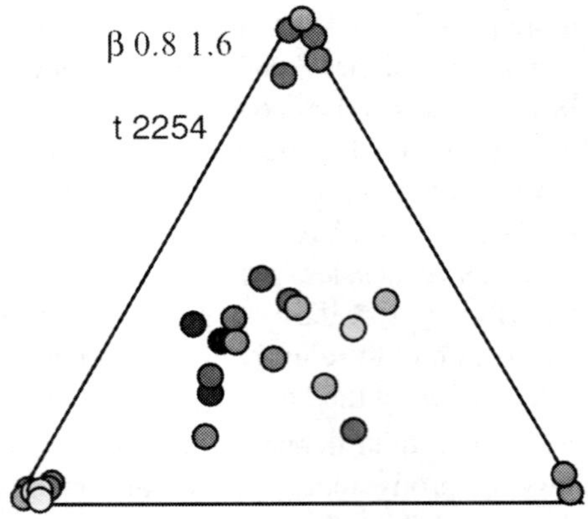

Fig. 15.4 The probabilities of buyers visiting the three sellers when buyers have high or low values of β.

Here some buyers have high β values and they remain at the apexes of the simplex whilst those with low β values remain in the middle of the simplex. The presence of the low β buyers does not prevent the high β buyers from becoming loyal

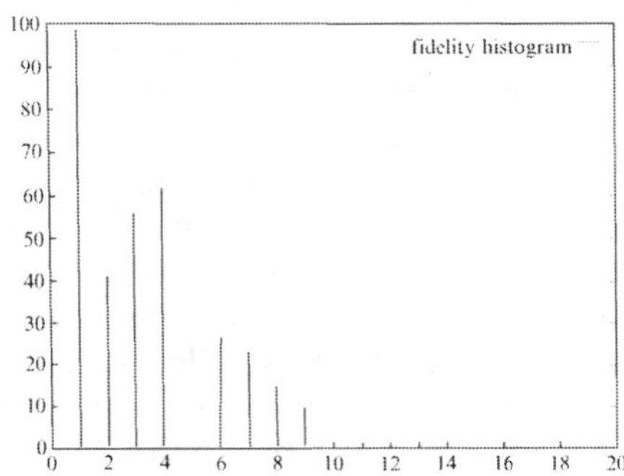

Fig. 15.5 The histogram of loyalties for buyers on the Marseille fish market, (source Weisbuch et al. (2000)).

The number of sellers from whom buyers bought more than 50% of their fish is plotted against the number of buyers. The distribution is clearly bimodal with most buyers buying most of their fish from one seller whilst the others buy most of their fish from several sellers with a mode of 4

from the fact that, for example, 48% of all buyers bought more than 95% of their cod from one seller, the seller of course, not being the same for all of these buyers. 33% of buyers bought more than 95% of their sole and 24% bought more than 95% of their whiting from one seller. In both the whiting and sole markets more than half of the buyers buy more than 80% from one seller. Furthermore, as the theory predicts, those sellers with the highest sales and those who come to the market most frequently are those who are most loyal. This recalls an earlier model of Whittle (1986) where there are two sorts of activities, farming and trading, and, under certain conditions, markets may emerge with individuals practising each of these activities where previously there were only itinerant traders.[4] Now, I would like to pursue this issue and point out that we can produce artificial markets in which

[4] A detailed discussion of this sort of problem is given by Aoki (1996), who thinks of the buyers in the markets as being partitioned between the sellers and each buyer as having a probability of transiting from one seller to another. He looks at the limit distributions of such a process.

individuals are faced with several periods and in which sellers modify their prices and quantities supplied in the light of past experience still reproduce the "stylised facts" of the simple theoretical model which are, in turn, consistent with empirical observations. Up to this point I have concentrated on one feature of real markets, that of trust and fidelity. This was because in the quotes mentioned earlier, this feature was described by practitioners as one which separates real markets from theoretical markets and what I am suggesting is that artificial markets can help to fill this gap. Now suppose that we are interested in reproducing other features of real markets such as dispersion of prices. How can we set about modelling artificial markets which will do this?

15.7 An Artificial Market Based on a Simpler Modelling Approach

As is clear from the previous sections the problem of modelling even such simple examples as fish markets is that if one tries to incorporate some of the realistic features of the microscopic interaction between the actors, the model rapidly becomes analytically intractable. We can, a as I did in the previous section, then simulate the theoretical model in more elaborate forms and check on the correspondence with the empirical facts. However, one may want to go beyond this sort of consideration and generate a model which is capable of producing several of the features of the real market. For this we have to start at a more basic level and construct an artificial market from the bottom up. One way to do this is that provided by the so called "multi agent" modelling approach (see for example, Arthur et al., 1997; Epstein, 2007). In this, agents are endowed with simple rules which govern their reaction to their economic circumstances. In the versions which interest me here, as time goes on, the agents place more weight on those rules which turn out to be more profitable. This shares the sort of updating from experience that I described in the previous models but one can generalise the rules to cover more general behaviour.

Thus, the previous model is a special case of this approach since the rule that the agents use can be thought of as determining the choice of seller in a particular way depending on the success of transactions with that seller in the past. However, once the rules are extended to cover a variety of choices the behaviour of the model quickly becomes too complicated to model formally even when drastically simplified.

Suppose then, that we wish to use the same sort of approach but want to analyse several features of the market at the same time. This suggests, by a sort of Occam's razor approach, simulating a model with as simple an updating procedure as possible. In such a model one hopes to find, as emergent features, some of the salient aspects of the real empirical markets that interest us. In Kirman and Vriend (2000) we developed a simple model which produces three of the features of the Marseille fish market. These are firstly, the division between loyalty and shopping behaviour on the part of buyers that we have already mentioned. Secondly, there is the price dispersion even for the same species. Lastly, we might hope that sellers learn to handle their clients in a way which corresponds to what actually happens in reality. In the theoretical model developed in Weisbuch et al. (2000) the problem of handling several periods in the day was already sufficient to oblige us to resort to simulations. Adding now the extra complication of pricing behaviour and the handling of clients means that there is little hope of producing a comprehensive theoretical model which will reproduce all the characteristics of the real market so we built a simple artificial market in which agents interact with each other and learn in so doing.

In the simple simulated model we developed in In Kirman and Vriend (2000), ten initially identical sellers and one hundred initially identical buyers met in the market hall for five thousand days for a morning and an afternoon session. They traded single individual units of a perishable commodity. Here we make two simplifications. The morning and afternoon sessions correspond to the idea that there are several opportunities to trade during the day. Taking two periods allows us to take account of the idea that the possibility of trading later in the day has an influence on the prices that buyers will accept and sellers will propose early in the day. It would, of course, be more realistic to consider more trading opportunities in the day. The single unit assumption is frequently used but can be criticised on the grounds that when buyers require different amounts this may influence what they pay. Leaving these caveats on one side for the moment I can now describe the model.

On each day the sequence of events in our artificial market is the following.

In the morning before the market opens the sellers purchase their supply outside the market for a given price that was identical for all sellers and constant through time. Thus, like the small open countries in trade theory, we assume that the participants on the Marseille have no influence on what happens in the outside world. The market opens and the buyers enter the market hall. Each buyer requires one unit of fish per day. All buyers simultaneously choose the queue of a seller. The sellers then handle these queues during the morning session. Once the sellers have supplied all the buyers who are willing to purchase from them the morning session ends. All those buyers who are still unsatisfied choose the queue of a seller in the afternoon. Sellers now sell to those buyers who are willing to purchase from them and the end of the afternoon session is then reached. All unsold stocks perish. Those buyers who did purchase fish resell that fish outside the market at a given price that is identical for all buyers and constant through time. Each buyer can visit at most one seller in the morning and one seller in the afternoon.

What are the decisions with which the actors are faced? Buyers have to choose a seller for the morning session. They then have to decide which prices to accept or reject during the morning session. If necessary, they also have to decide on a seller for the afternoon. Lastly, they must decide which prices to accept or reject during the afternoon session. Sellers have also four decisions to make. They must decide what quantity to supply. They must decide how to handle the queues with which they are faced. They must decide which prices to set during the morning session and which prices to set during the afternoon session.

condition	action	strength
if	then	....
.. ..	....	....
.. ..	....	......

Fig. 15.6 A classifier system, (source Kirman and Vriend (2000)).
Each rule has a condition that has to be satisfied, an action to be taken and a strength attached to it

In the model described each individual agent uses a Classifier System for each decision and this means that each agent has four such systems "in his head." A simple stylised classifier system is presented in Fig. 15.6 which is taken from Kirman and Vriend (2000).

Each classifier system consists of a set of rules. Each rule consists of a condition "if......." and an action "then......" and in addition each rule is assigned a certain strength. The classifier system decides which of the rules will be the active rule at a given point in time. It checks the conditional part of the rule and decides amongst all of those rules for whom the condition is satisfied which to choose. This is done by a simple auction procedure. Each rule makes a "bid" to be the current rule and this bid = current strength + ε, where ε is white noise, a normal random variable with mean 0 and fixed variance. The rule with the highest "bid" in this auction becomes the active rule. The white noise means that there was always some experimenting going on and there was always some probability that a rule, however bad, will be chosen. At time t the classifier system updates the strength s_t of a rule that has been active and has generated a reward at time $t-1$ as follows:

$$s_t = s_{t-1} - c \cdot s_{t-1} + c \cdot reward_{t-1}, \text{ where } 0 < c < 1. \qquad (15.12)$$

Hence, as long as the reward generated by the rule on day $t-1$ is greater than its strength at $t-1$ the strength will increase. The strength of each rule converges to the weighted average of the rewards generated by that rule. What the reward is, will depend on the rule in question. Supposing that in our market example the rule for the buyer is of the form, "*if the price proposed by the seller for one unit of fish in the morning is 11 euros then accept.*" The reward for using this rule would then be the profit that is generated by using it. In this case the reward would be the price at which the unit of fish is sold on the retail market minus the price paid (11 euros). When the model is started the strengths of all rules are equal.

What the agents in this model are doing is learning by an even simpler version of reinforcement learning than that encountered previously. Details of the particular rules in the simulation model of the Marseille fish market can be found in Kirman and Vriend (2000).

Although such an approach seems to be innocent of theoretical pre-suppositions it should be noted that the very choice of the rules amongst which the agent chooses has an impact on the outcomes. Ideally, one would like to start with agents who are totally ignorant. However, this would imply that they would somehow generate a set of rules with which they would experiment. This pushes the analysis back many stages to a very fundamental level. What is done here is in line with standard practice which is to provide the agents with a set of rules and simply note that this, to some extent, conditions the outcomes of the process. As an example, consider the fact that we would like agents to learn how to handle the queues with which they are faced. In an ideal world we would like the agents to realise that their handling of the queues is important and then for them to work out how to handle them. As it is, by giving different rules explaining how to handle queues the modeller is already biasing the behaviour of the seller by suggesting to him what it is that is important in generating his profit. However, what is not biased is the choice amongst the rules presented. Thus, the rule chosen will be the best available for handling queues amongst those presented, given the agent's experience, but he might well himself have focused on some other aspect of the market. With these reservations it is still worth examining the results of the simulations and to see to what extent they reflect reality.

One might ask whether some of the features of the market could not be modelled in a more theoretical way. For example, in the last session it seems as if buyers and sellers are faced with a version of the ultimatum game (see e.g., Güth and Tietz, 1990). Since, in the model sellers propose a price and buyers either accept or refuse it would seem that the sensible price to propose is just slightly less than the price at which the fish can be sold on the outside market and this indeed is the sub-game perfect outcome of the ultimatum game. However, it has long been noted that this is not what one observes in experimental outcomes nor in reality. One obvious reason for this is that we are observing a repeated game in the market. Thus, a refusal today has implications for behaviour tomorrow even if agents are not aware of this. Buyers learn to accept or reject prices on the basis of the profitability of doing so whilst sellers learn in a similar way which prices to ask. What is crucial here as noted by Roth and Erev (1995) is that the relative speed of learning on each side of the market will govern which outcomes occur. The importance of this will become clear as soon as we look at the results of the simulations. However, once again it should be emphasised that the agents in the model are not behaving strategically, they are simply learning from experience. It may be the case that they are led to behave in the same way as if they had reasoned in a game theoretic way but this would be an emergent phenomenon and not one which was built in to the models of individual reasoning.

In the simple artificial market let us first look at the prices asked and accepted in the morning as shown in Fig. 15.7.

There is first of all a considerable period during which learning takes place and then prices settle to 10 which is one greater than the price at which fish is bought by sellers outside the market and one greater than the perfectly competitive price. What is interesting is that during the learning period, which is what governs the final outcome, two things are going on. Sellers learn to ask prices close to the ultimatum

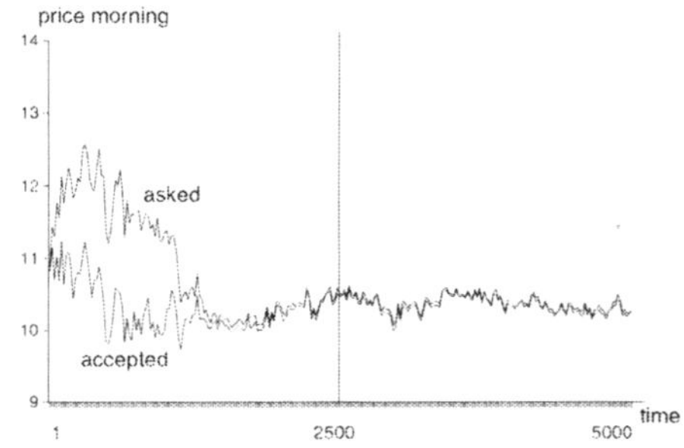

Fig. 15.7 Prices asked and accepted in the morning session, (source Kirman and Vriend (2000)).

Note that when learning has taken place the two graphs converge

price which is 14 euros, one less than the price at which fish can be sold on the outside market. However, buyers do not learn as quickly to accept such prices. Where does this difference come from? The answer is that initially some buyers will accept high prices having not learned to do otherwise. This will encourage sellers to charge such prices. However buyers will start to find out that they can obtain higher profits by refusing high prices and accepting lower ones. There are always some such prices to be found. As sellers learn that buyers are not accepting their prices they start to decrease the prices asked and simultaneously as buyers observe that prices being asked are descending they start to decrease their acceptance levels. Once again, sellers learning "leads" that of buyers and as a result the prices converge. In the afternoon, there are also two separate learning processes going on and once again convergence occurs but to a higher price (11euros) than in the morning.

This might seem extraordinary since if buyers become aware that prices in the afternoon are higher than prices in the morning they should presumably always buy in the morning. This is not correct. The reason is simple, those people who reappear in the afternoon have been selected. To see this, consider a situation in which the distribution of prices asked is the same in the morning as in the afternoon. Suppose now that those buyers that encounter prices in the upper tail of the distribution reject in the morning, the result of this would be that the average price paid in the morning will be lower than the average price paid in the afternoon. The whole point here is that it is not the average price that is rejected whereas what is shown in the figures is the average at any point in time. In Fig. 15.8 the price distribution over the last 2,500 days is shown and it can be seen that it does not become degenerate and concentrated on one price.

Thus, a phenomenon which is observed on the real market in Marseille, as we saw earlier, emerges in our artificial fish market.

A second feature of the real market is that which has been discussed earlier, that of "loyalty." In the previous model we simply established the pattern of loyalty but did not suggest any macroeconomic consequences of that feature. To pursue this question we need to have some measure of loyalty and then to examine its impact. To do this we construct an index of loyalty which has a value equal to one if the buyer is perfectly loyal to one seller and has a value equal to $1/n$ where n is the number of sellers when the buyer systematically visits each seller in turn, that is, when he has the most extreme "shopping around" behaviour. More specifically, the

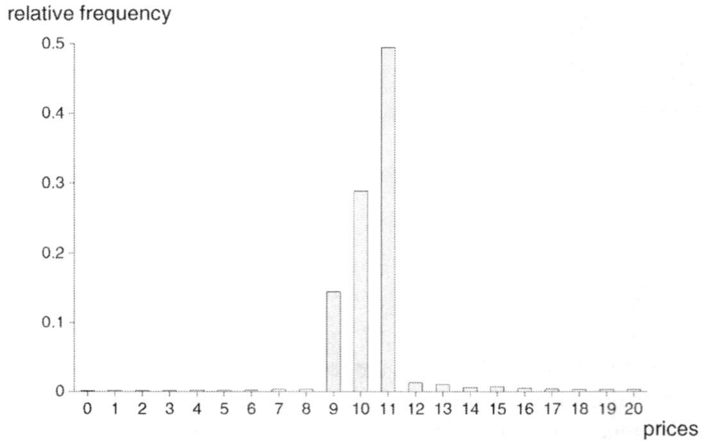

Fig. 15.8 The distribution of prices, (source Kirman and Vriend (2000)).
What is shown here is the average of the distribution of prices for each of the last 2,500 days

loyalty index is given by:

$$L_{ij}(t) = \sum_{x=1}^{t} \frac{r_{ij}(t-x)}{(1+\alpha)^{t-x}} \tag{15.13}$$

This is an indicator of how often buyer i visits seller j. It is a global statistic covering the whole period but there is a discount factor represented by α. The parameter $r_{ij}(t)$ is a counter which increases with each visit of i to j. Here we took $\alpha = 0.25$ and $r_{ij}(t) = 0.25$ if buyer i visits seller j at time t and $= 0$ otherwise.

In the spirit of this approach, nothing was built into the rules of the sellers to make them privilege loyal buyers. We wanted to see whether this sort of behaviour would emerge. The sort of rules they had were of the form:

"If loyalty = a certain value then choose a certain probablity of serving that client",
"If loyalty = a certain value then charge p"

What probability will be chosen depends on whether the seller learns to favour loyal customers or not. Which p is charged depends on how successful that choice turns out to be. The time series of average loyalty is shown in Fig. 15.9.

What happens is that 90% of the buyers actually get a higher pay off by being loyal as can be seen in Fig. 15.10. What this means is that when basically loyal customers shop around, as they do stochastically from time to time, the profit realised is lower on average than when they buy from their regular supplier.

Furthermore, nine out of ten of the sellers get a higher profit when dealing with loyal buyers as shown in Fig. 15.11. In other words the profit, on average from a loyal customer is higher than from a random shopper. Here the difference in revenue represents the fraction of the average revenue from loyal customers above or below the average profit realised from transactions with casual buyers.

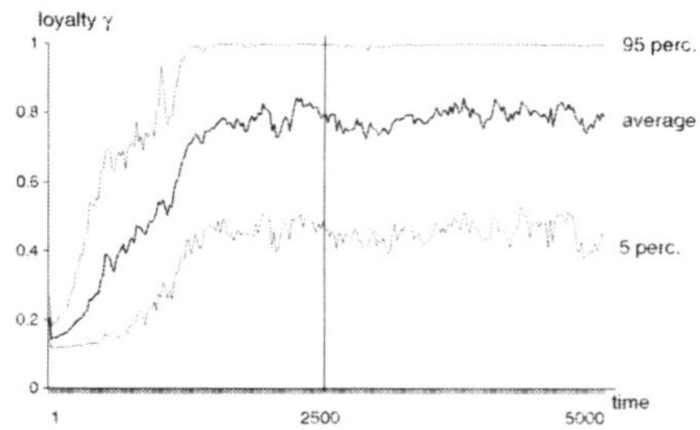

Fig. 15.9 The evolution of loyalty, (source Kirman and Vriend (2000)).

Average loyalty and 95% and 5% limits for loyalty over 5000 periods

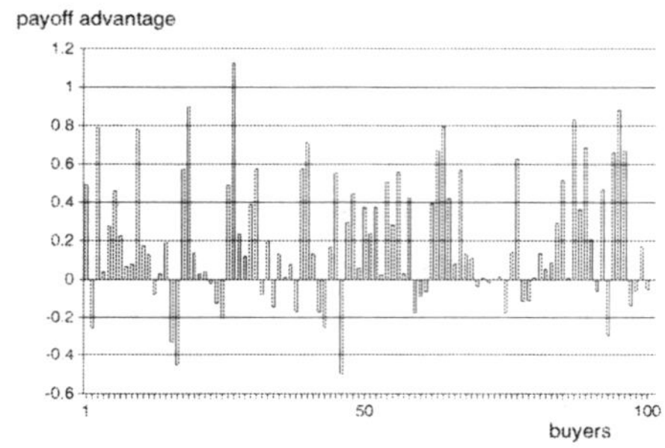

Fig. 15.10 The pay-off advantage for buyers of loyalty, (source Kirman and Vriend (2000)).

The advantage in terms of pay-off achieved when being loyal as opposed to "shopping around" for each of the 100 buyers

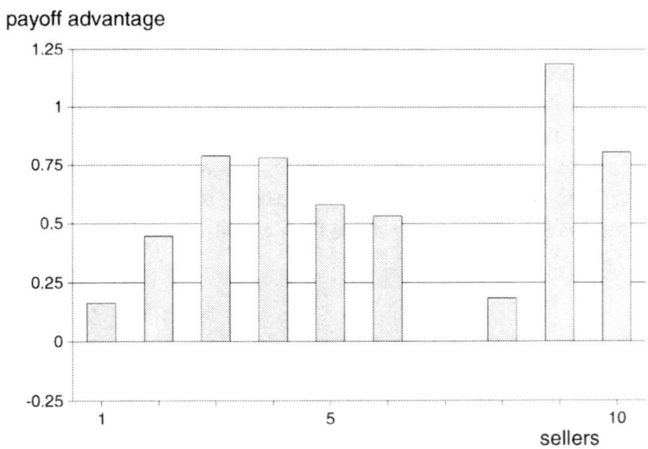

Fig. 15.11 The pay-off advantage for sellers of loyalty, (source Kirman and Vriend (2000)). All sellers but 1 achieved a higher pay-off from their loyal customers

This is a reflection of the fact that what is happening here is not a zero sum game. Only when a transaction takes place do buyers and sellers realise a profit. Thus, payoffs will be highly conditioned on acceptance and rejection, and on prices asked. The question then becomes how do loyal buyers tend to be handled by sellers? In all but one of the cases, sellers learn to give priority in service to loyal buyers but to charge them higher prices than random shoppers. Buyers learn that when they become loyal their profit is higher since they are more likely to be served even though they pay higher prices. Thus, loyalty is profitable both to buyers and sellers.

What about the one seller who did not find loyal customers more profitable than shoppers? This seller learned to charge low prices to loyal customers but to give them low priority in the queue. One might ask why he did not learn to adopt the more profitable strategy learned by the other sellers. The answer here is that with the sort of local learning that is going on a move towards better service and higher prices for loyal customers can never develop. To make such a move would imply increasing prices and improving service. However, buyers will immediately observe the higher prices and will not necessarily immediately observe better service in terms of priority in the queue. This will lead them to reduce their probability of visiting this seller. As this seller observes that his customers are drifting away he will go back to his former behaviour and will therefore never learn his way to the more profitable strategy. However, it is interesting that, in the model this seller still makes profits so he does not disappear. Thus there is at least one explanation for the dispersion of profits that one observes on the market. No figures are available to document this but there is a consensus on the market that some sellers make considerably more profit than others and the argument of our model would be that they have simply reinforced on more profitable rules.

This very simple rudimentary artificial fish market model manages then to reproduce some of the features of the real market. For example, it is interesting to note that on average in the Marseille fish market loyal buyers pay higher prices than shoppers. Those buyers who buy more than 50% of their fish per year from one seller pay, on average 5% more than the other buyers even though almost all the large buyers are loyal. Thus here we have a clear organisational feature which has emerged and which has had a very specific consequence for the market price distribution. Such a model has the advantage that it can always be extended to examine other aspects of the real market whereas to attempt to construct a theoretical model which incorporates all of these is a more than ambitious task.

15.8 Other Forms of Market Organisation

An important reason for constructing artificial markets is that it is possible to incorporate the particular structure of a real market and then to study what happens when that structure is modified. We have studied a number of markets for the same product which have different organisations and then have looked at the outcomes on each to see if the organisational features have a real impact on the aggregate market behaviour. I have discussed at some length the features of the wholesale fish market in Marseille where all the activity takes place through pairwise trading.

As a comparison we have analysed the fish market in Ancona, on the Adriatic coast of Italy, which is organised as three simultaneous Dutch, (descending price) auctions, (see Gallegati et al., 2007). We have detailed data for all the transactions made on this market and this provides us with an opportunity to compare the data from this market with that in Marseille and see if similar stylized facts emerge. Secondly we also have the possibility to see if our data exhibits the features found in auction data or predicted by auction theory. Before describing the artificial market model which we constructed let me start by giving a brief description of the real Ancona fish market which is known as MERITAN.

15.9 MERITAN a Market Based on Dutch Auctions

The MER.IT.AN ("MERcato ITtico ANcona" Italian for Fish market of Ancona) is open 4 days a week (Tu.-Fr.; 3.30–7.30). It consists of 3 simultaneous Dutch auctions with about 15 transactions in total per minute. The total value of the fish sold amounts to 25million euros per year. Each type of fish is arranged in cases of about 5–7 Kg. Each morning the vessels are randomly assigned to one of the three conveyor belts and then all the cases from one vessel are put on the belt assigned to it. When the selected seller puts a case on the belt the price display is set (the auctioneer decides the initial price) and starts going down while the case moves toward the end of the belt. Buyers watch the three displays and can bid on one or more of them, the first person to push the button at the price that has been reached wins the auction.There are about 170 buyers. 20 of them are wholesalers while 150 retailers (ambulant sellers, outdoor market stallholders, fish shops); in any case they are not the final consumers. There are 70 sellers.

The data we used correspond to the only one of the three conveyor belts for which the information was collected electronically at the time. They cover the period from the 19th September 2002 to the 28th of May 2003. The database represents 53,555 transactions for a total weight of 360,115 kg. During this period 70 sellers and 149 buyers exchanged fish of 110 transaction classes on this specific conveyor belt (data for the whole market are more comprehensive). Note that what we call a transaction class is different from a species (see the second column of the table below for examples of transaction classes).

The data are collected daily on the market computer. For each case traded the data provides:

- The day, month and time (hour and minute) of the transaction.
- The weight.
- The price per kilo and total.
- The identification number of the seller (vessel).
- The identification number of the buyer.
- The transaction class (T.C. hereafter) and its identification number.

To complete the description of the market we examine buyers and sellers size distributions. We define the size as the total weight of the fish bought or sold by the agents over the period. As shown in Figs. 15.12 and 15.13, there is no dominant size among the sellers while the buyers are clustered on the small size. The distribution of buyers presents a notable peak while that of sellers is rather flat. On the other

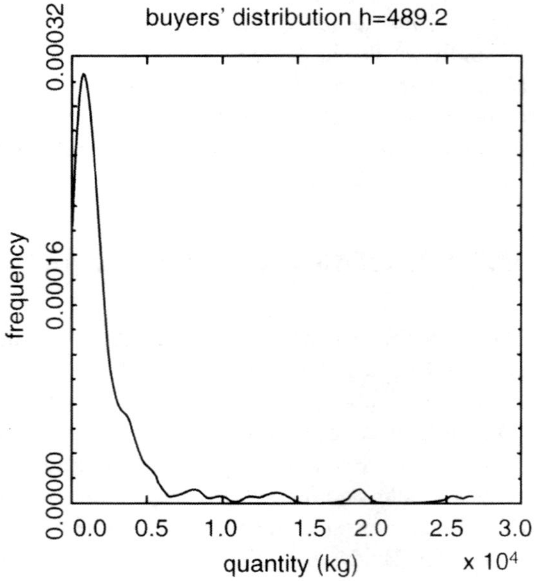

Fig. 15.12 The distribution of the total amounts bought on the Ancona Market, (source Gallegati et al. (2007)).

This is a non-parametric fit of the histogram of the quantities purchased over the whole period, using kernel smoothing techniques with a Gaussian kernel, where h is the bandwidth

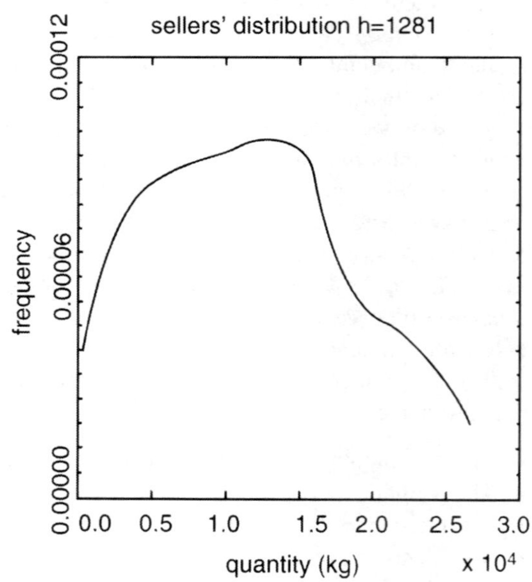

Fig. 15.13 The distribution of the total amounts sold on the Ancona Market, (source Gallegati et al. (2007)).

This is a non-parametric fit of the histogram of the quantities sold over the whole period, using kernel smoothing techniques with a Gaussian kernel, where h is the bandwidth

hand it is evident that there are a few very large buyers. Here, as in Marseille, we are dealing with a heterogeneous group of buyers and sellers but this time they interact through a central auction mechanism. The buyers observe who purchased which lot and can therefore, in principle, act strategically. However, since there is a transaction every 4 s, it would seem more plausible that the buyers use rules of thumb rather than detailed and complicated calculations. Again we cannot rule out the possibility that the two have come to coincide as a result of learning over a long period. Some anecdotal evidence that buyers are not able to calculate their optimal strategy is provided by the auctioneer at Sete in the South of France, a fish market which also operates on a Dutch auction basis. I observed that sometimes he started the auction at a price which was immediately taken and that this might indicate that the lot would have been taken at a higher price. He admitted that this was true but said that he did this to speed things up and to stop the buyers thinking strategically. In this way he claimed that he was increasing his total revenue. He gave, as an example, the card game, bridge and pointed out that if one played relatively slowly one had time to reflect on which cards had passed while this becomes impossible if one is forced to play quickly.

The thing to note here is that we are dealing with a distinctly heterogeneous population both of buyers and sellers which makes a complete formal analysis very difficult, if not impossible.

15.10 The Empirical Evidence

We took a closer look at two questions which, we believe might help us to understand whether the market mechanism does have an impact on market behaviour and we then built an artificial market model to see whether we could reproduce the specific features of this type of market.

First, we analysed the way in which prices are formed and the dynamics of this process.

Second we investigated the effect of the auction mechanism. The two main questions here are:

- Does the auction destroy buyer-seller relationships?
- Despite the auction mechanisms are there buyers (sellers) who systematically obtain lower (higher) prices than others?

15.11 Price Dynamics

The first question that we asked and one which has a long history in economics, is how do the prices evolve over the day? To analyse the price dynamics during the day we developed two types of graph. We first ranked the daily transactions by the time of day in which they occurred and then we performed averages for the transaction with the same rank. As shown in Fig. 15.14 the average price goes down as the rank of the transactions increases. A strange feature appears which we also observed for certain species of fish on the Marseille market: For a large number of T.C.s the average price starts increasing for the last transactions. Without knowing the exact information available to bidders it is difficult to explain this. However, if certain buyers need certain quantities of particular fish and suspect that the market is coming to an end then this could make them increase their bids.

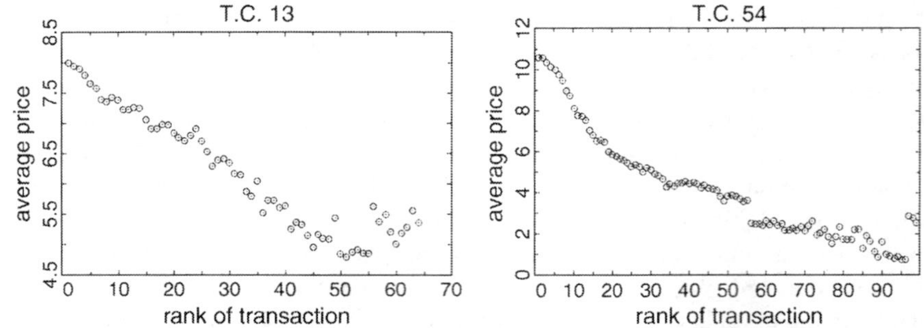

Fig. 15.14 Average price for each rank of transaction (the first dot on the left, for instance, is obtained collecting the price of the first transaction for each day and computing the average). (source Gallegati et al. (2007))

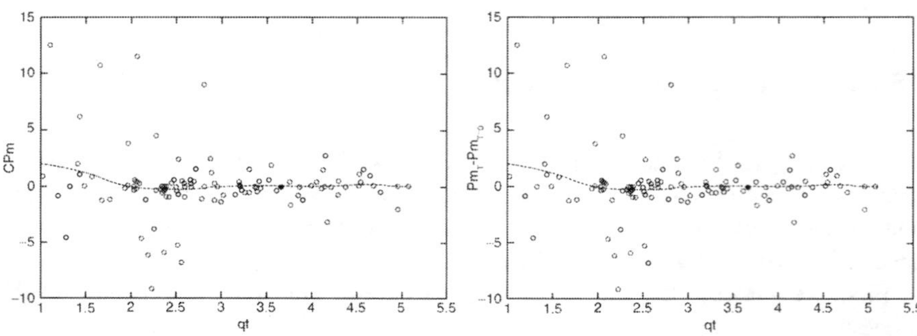

Fig. 15.15 Scatter plot of price variation for the last two transactions, PmT -PmT-2 versus daily quantity of fish qt. The left panel show a piecewise linear regression using the threshold qt = 2.8 whereas the right panel shows a non-parametric regression with Gaussian kernel and bandwidth 0.4. (source Gallegati et al. (2007))

The second type of plot for the analysis of price dynamics is shown in Fig. 15.15. Here we plot the data corresponding to the day with the largest number of transactions for the specific T.C. (the horizontal axis records the time of the transaction so a value of 4.5 means that the transaction took place at half past 4 in the morning). It is evident in general that the price volatility decreases as the auction proceeds. The early prices exhibit a turbulence that disappears as the auction proceeds. This is probably due to two things. Firstly at the outset there are buyers that have to buy very early in the morning for particular reasons (they may have a rather long trip back to the place where they resell the fish). Such individuals do not have the possibility of learning about the evolution of prices. The second reason is that there is learning going on and buyers are seeing how much fish of a certain type has passed and the prices of the transactions.

Another question is the dependence of the price level on the day of the week. One might expect that, for example, fish would fetch better prices on Fridays for purely historical reasons, since in catholic countries families still tend to eat fish on Fridays. However, we found that, if anything, prices decline somewhat over the week. The Friday effect disappeared long ago in the U.K. for as Lionel Robbins (1935) remarked about the reformation in the U.K which replaced catholicism with anglicanism, "The influence of the Reformation made no change in the forces of gravity. But it certainly must have changed the demand for fish on Fridays."

15.12 Loyalty Again

A very interesting question is to ascertain whether the auction mechanism destroys buyer–seller relationships. As I have mentioned, in Weisbuch et al. (2000) we found strong evidence for loyalty of certain buyers to sellers in the Marseille fish market. But there the market is characterised by bilateral bargaining and the buyer could, in principle, collect every sellers price to choose his best action. However many do not do so and as we saw loyal buyers actually pay more than random shoppers. Now the question is what does loyalty mean in an auction market. Remember that the name of the vessel which caught the fish is posted as the fish arrives on the band. So, it is possible that some buyers come to appreciate the fish from particular boats. Since the buyers are not the consumers they presumably realise greater profits from this fish. It is therefore possible that we see loyalty emerge. This does not mean that there is no uniform basis for judging the quality of fish. Supposing one vessel has the "best" fish. Then the prices for the fish from that vessel will be bid up and this may exclude a number of buyers. Thus we may see buyers becoming loyal to different buyers and paying different prices as a consequence. In fact, Buyers in Ancona learn to become loyal as in Marseille but the pattern is somewhat different. What sort of measure should we use to calculate the extent of loyalty? A typical measure is the Gini index which indicates how spread out among different sellers are the purchases of a particular buyer. We calculate the Gini index for each buyer. In Fig. 15.16 we show the Lorenz curve for the two extreme cases (the least concentrated in the left panel and the most concentrated in the right one).

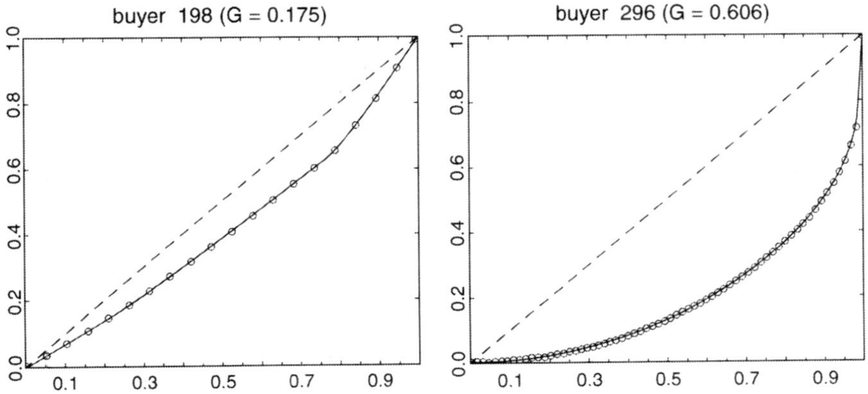

Fig. 15.16 Lorenz curves for the two most extreme buyers. (source Gallegati et al. (2007))

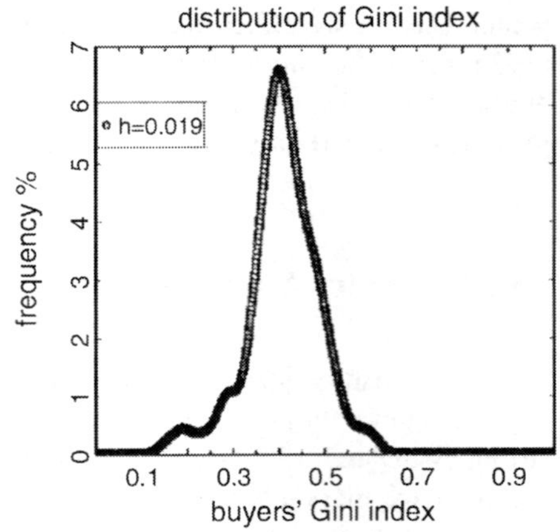

Fig. 15.17 Frequency distribution of the Gini indices of all buyers. (source Gallegati et al. (2007))

To have a global picture of the market we made a smoothed (Gaussian kernel) frequency distribution of the Gini index among buyers. A significant share of buyers have a Gini index equal to 0.4 and almost all have their index between 0.35 and 0.55. As I have said the buyer–seller relationship is different from that in the Marseille fish market. There, (Weisbuch et al., 2000) we found basically two type of agents: those who were totally loyal (these would have a Gini coefficient of 1) and those that did not care whom they buy from, (these would have a Gini coefficient close to 0). The auction mechanism washes out this distinction since the distribution in Fig. 15.17 is single peaked. However, we do see some preference on the part of different buyers for the vessels whose fish they buy. Different buyers are loyal to different vessels even though this loyalty is far from total. Can we say more about loyalty? An obvious question is as to whether loyalty depends on the buyers size. We found that the amount of loyalty increases with the size of the buyers up to a given value. Beyond this level the concentration index decreases or stays stable. This is probably because very large buyers are forced to neglect the source of the fish if they are to get all the fish they need. Before looking at what prices different buyers and sellers pay, which is linked to the loyalty problem, one remark is in order. Vessels are of different sizes therefore if there was one very large vessel and many small ones, for example. It could be the case that many buyers buy a disproportionate amount of their fish from that one vessel. Thus loyalty would be a reflection of the size of the vessels. It could also be that different vessels manage to obtain very different prices which could sort the buyers into classes. With this in mind we looked at the prices that buyers pay and sellers obtain.

A first observation is that it seems that the amount of fish bought or sold has almost no influence on the prices associated with the buyers while for sellers we observed that the vessels with larger catches never sold at an average price lower than 7 euros, while some of the smaller sold at lower average price. As on the Marseille market we do not observe classic individual demands, but we do observe different individuals paying different prices for the same fish. Nevertheless, what we see here is once again the diversity of behaviour at the individual level and some rather coherent aggregate structure. The market can be considered as an algorithm which transforms the varying individual demands into prices and organises itself so that all the fish is sold. As one might hope higher average prices are obtained on days when supply is shorter but this is not a reflection of classical individual demands. There are individuals who pay, on average, higher prices than others but sometimes these individuals beat their colleagues and obtain fish at low prices. In the same way it is not true that the prices of fish from those vessels that realise higher average prices always dominate those of their lower price competitors. There are many explanations for this and, for example, a buyer who normally pays high price may obtain almost all of his requirement early in the day and be prepared to wait and to try to complete his quota at a lower price.

15.13 Comparison Between Auctions and the Decentralised Market in an Agent-Based Model

Now comes the question as to the difference between the two market systems. There is a theoretical literature on this subject but it is limited to very restrictive models where one can hope to derive analytical results. The usual preoccupation in this type of model is with the efficiency of the outcomes, where an outcome is considered as efficient if it maximises the sum of sellers and buyers pay-offs. The typical basis for the analysis of the auction market is the so-called "private values" model in which agent has his own reservation price for the object to be auctioned and this is drawn from a known distribution but unknown to the other buyers. Myerson and Satterthwaite (1983) showed that when values are private then the result of the negotiation process will be inefficient in the sense just defined. Bulow and Klemperer (1996) show that under certain assumptions, if the buyers values are independent then the auction mechanism dominates pairwise negotiation. Xiaohua and McAfee (1996) show that both mechanisms can produce efficient equilibria but that those associated with pairwise negotiation are not evolutionarily stable. Finally Manelli and Vincent (1995) show that when goods can have different qualities negotiation with sequential offers can give a more efficient outcome than an auction.

In the experimental literature Roth and Erev (1995) compared two mechanisms in different countries. In their experiments the negotiation corresponded to the "ultimatum game" which is rather different from the mechanism we are studying here. They compared this to an auction and found that contrary to theoretical predictions, in both cases the result was efficient and that the seller obtained almost all of the surplus. A rather strange result was obtained by Pogrebna (2006) in which she found that sellers did better in bilateral bargaining than at auction.

Faced with these results, this was the obvious situation in which to use artificial markets. In Kirman and Moulet (2008) what we do is to build agent based models which allowed us to make as direct a comparison as possible between MERITAN and Marseille. I shall only sketch the outline of the models here but the full details can be found in Kirman and Moulet (2008). In order to do this we modelled both types of market giving the corresponding rules to the participants.

15.14 Common Features

In both markets we consider that the buyers have an outside market on which they are local monopolists and to simplify matters we assume that demand is of the form,

$$D_i(p) = q_i = a_i - b_i p \tag{15.14}$$

and since the buyer has already purchased his stock on the wholesale market this determines the price at which he can sell and his profit.

The buyer can calculate the value of buying an incremental unit over and above his n previous units which is simply $\pi_t^{n+1} - \pi_t^n$ where,

$$\pi_t^n = n\frac{a-n}{b} - \sum_{k=1...n} p_k \tag{15.15}$$

where p_k is the price paid for the kth unit. This gives us the incremental profit which is

$$\delta\pi \equiv \pi_t^{n+1} - \pi_t^n = \frac{a-(2n+1)}{b} - p_{n+1}. \tag{15.16}$$

The buyer will then leave the market if at no price will the incremental profit be positive. That is when,

$$\delta\pi < 0 \Leftrightarrow n > \frac{a-1}{2}. \tag{15.17}$$

This reasoning applies whether the buyer is in the decentralised market or participating in an auction. Note that in giving these conditions we attribute considerable rationality to the buyers. In the "artificial life" approach the buyer would have to learn what was the appropriate rule to follow. It is always a delicate issue in modelling artificial markets to decide what level of basic "rationality" one wishes to attribute to the actors. If one starts from a level of total ignorance the learning process may take arbitrarily long, however assuming that agents behave in a completely rational way undermines one of the major reasons for undertaking this sort of exercise.

What we will see is that whenever an agent uses a rule, for example, proposes a certain price, or decides upon a bid at a certain level this will generate a certain profit and this will be used to update the weight of that rule. Thus if we consider $J_{ij}(t)$ as the total profit obtained by buyer i when using the rule j up until time t and $\pi_{ij}(t)$ as the profit generated by that rule at time t then as in our earlier model we update as follows,

$$J_{ij}(t) = (1-\gamma)J_{ij}(t-1) + \pi_{ij}(t). \tag{15.18}$$

These profits are transformed into the probabilities of using the rule in the same way as before, that is,

$$p_{ij}(t) = \frac{e^{\beta J_{ij}(t)}}{\sum_k e^{\beta J_{ik}(t)}}. \tag{15.19}$$

Now all that remains is to specify the rules employed by the actors in the two different mechanisms that we wish to study.

15.15 The Auction Market

The agents rules will be of the conditional type, as used in the classifier system in Kirman and Vriend (2000). The seller, that is the owner of the vessel fixes the minimum price that he is prepared to accept and the auctioneer then decides on the starting price for the auction which obviously has to be above the minimum decided by the seller. So the decisions are of the form:

Open the auction at $P_{max} = x$.

Similarly the seller has a rule of the form

Withdraw the lot at $P_{min} = x$

where x takes on integer values from 0 to $\bar{x}$.

Buyers Each buyer has a reservation price Pr which is the maximum that he is prepared to pay for the lot. What he does is conditional on the decisions of the seller and the auctioneer. He thus has rules of the following type:

If the opening price $P_{max} > P_r$ *then bid at* P_r,
If the opening price $P_{max} < P_r$ *then bid at* P_{max}.

What profit should one associate then to each rule? The buyer i updates his rules at the end of each market day and takes account of the amount of goods q_i that he actually bought and the quantities that he would have been able to buy. We assume here, for simplicity that one unit is sold at each round of the auction.

Given the demand that he faces on his retail market he knows the price that he could charge had he bought the optimal quantity and we denote this by p_{opt} and his aim is to acquire the associated quantity. There are three cases,

(A) If the buyer bought exactly the quantity he wished for then he knows that if he had bid at a higher price he would also have obtained that quantity but his profit would have been lower. However, if he had bid at a lower level then he does not know if he would have obtained the lot. If then we denote by p the price of the transaction, he only knows the profit he would have obtained if his bid had been $x \geq p$ in which case $\pi = p_{opt} - x$.

(B) If the buyer did not make a purchase he knows at what price all of the goods were actually sold. He can use this information to update his rules. Again he has, as his aim, to buy the optimal quantity q_{op} to obtain the optimal price p_{op} on his retail market. In particular since he wished to buy q_{op} he can observe the q_{op} sales at which the price was the lowest. He knows that had he bid above the winning price at these auctions he would have obtained the goods and his profit from each winning bid x that he made would have been $\pi = p_{opt} - x$. Of course had he decided on a bid below the selling price he would have obtained $\pi = 0$.

(C) If the buyer obtained a quantity of goods below his optimum then he can combine the two types of inference in (A) and (B).

Given this we know how to attribute profits to the rules used by the buyers at auction and, in particular, to some of those rules which were not used.

Sellers Suppose the seller proposes to withdraw his good at $P_{min} = y$, but it is sold at $x > y$, then we can write the gain from the rule set $P_{min} = y$:

$$\pi = \{x \text{ } if \text{ } y \leq x \text{ } and \text{ } 0 \text{ } if \text{ } not\}$$

where x is the winning bid. Obviously if there was no bid above y the good is not sold.

The auctioneer If the auctioneer wishes to evaluate the rule *Open the auction at* $P_{max} = y$ he does so as follows.

If he did not the rule and the good was actually sold at a price x higher than y then his profit had he used that rule would have been y and so we have

$$\pi = \{y \text{ } if \text{ } x \geq y \text{ } and \text{ } x \text{ } if \text{ } not\}$$

However if the auctioneer uses the rule, Open the auction at $P_{max} = x$, but the good is withdrawn by the seller at z then the profit from that rule will obviously be evaluated at $\pi = 0 \text{ } if \text{ } z \geq x$.

Having specified the rules, the pay-offs from them, and the resultant probabilities of using each of the rules the auction market can be simulated.

Pairwise trading First of all one has to specify the protocol for the negotiation and the following seems to be reasonably representative of what happens on the market in Marseille.

1. Seller j proposes a price p_1 and calculates or chooses a minimum price he is willing to accept, p_2
2. Buyer i accepts or proposes a price p with $p \leq p_1$
3. If $p < p_2$ then there is no transaction and if $p \geq p_2$ then there is a transaction at p.

Buyers The price p proposed by a buyer i is conditional on the seller j with whom he is faced and are of the form:

If the seller j offers a good then make an offer at x.

Sellers The seller decides on the first price to offer p_1 and on the minimal price p_2 which he is prepared to accept from the buyer in question, so the rule is conditioned on the buyer and are of the form:

If the buyer is i then make an opening offer of $p_1 = x$.

If the buyer is i then do not accept any counter offer $p_2 < x$.

15.16 Profit Generated by the Rules

Buyer If the buyer completes a transaction with a seller at the price p he knows that he would also completed the transaction had his offer x been higher i.e. $x > p$ but he would have had a lower profit which he can calculate simply as $\pi = p_{opt} - x$ so he can update the profit from all the corresponding rules. However, he cannot update the rules involving lower prices. If the seller refuses an offer p then the buyer can conclude that had he made a lower offer that too would have been refused generating a profit of $\pi = 0$.

Seller

(A) Suppose that the seller proposed as his opening offer p_1. If the resultant transaction was at the price p, the seller knows that for any opening price $p_1 \geq p$, the profit would have been $\pi = p$. But for any opening price $p_1 < p$, the profit would have been $\pi = p_1$ for the buyer would have accepted p_1. If the good was not sold then $\pi = 0$.

(B) If the seller chooses a closing price p_2, that is the minimum he is prepared to accept and the good was sold at p, then the seller knows that for any $p_2 > p$, the profit would have been $\pi = 0$. However of any choice $p_2 \leq p$, the profit would have been $\pi = p$. If the good was not sold when he made the choice of p_2 then $\pi = 0$ and the same would have been true for any $p > p_2$.

Differences between the two mechanisms The two mechanisms differ in the order in which actions are taken and in the information that the participants have. In the auction situation all buyers know at what price goods were sold or taken off the market so they compete directly with each other and the information used to update the rules is the same for every participant. However, in pairwise trading the prices that are proposed and the results are only known to the two individuals directly involved and the agents have no access to information about other transactions. Their updating of their rules is based only on their own transactions. Furthermore the buyer in a pairwise negotiation does not know the reservation price of the seller.

15.17 Simulations

All the simulations were run for a period of 20,000 "days." Both the auction mechanism and the bilateral trading framework were tested in each case.

We wished to evaluate the profits obtained by the agents as well as the quantities transacted. We tested situations in which only one side of the market learned and ones in which both learned. We also looked at two situations for the buyers, a homgeneous one in which they all faced the same demand on their retail market and a heterogeneous one in which there were two groups of buyers facing different demand curves. We excluded the uninteresting case where there was excess supply at all prices.

We considered, as already mentioned the buyers as all facing a linear demand curve at home of the form

$$D_i(p) = q_i = a_i - b_i p. \tag{15.20}$$

In the homogeneous case all buyers faced a home demand of

$$D(p) = q = 6 - 2p. \tag{15.21}$$

In this case they always leave the market as soon as they have bought 3 units of good.

In the heterogeneous case the buyers differ in the demand they face at home and we capture this by simulating an example with three buyers who have respectively $b_i = 1, 2$ and 3 and all of whom have $a_i = 6$.

Once we fixed the characteristics of the agents we tested two values for the total supply. In the first case, given that we have three buyers each of whom would optimally buy 3 units we set the global supply $Q_0 = 9$.

In the second case we set $Q_0 = 6$, which means that there were three units less than the optimal quantity desired by the buyers.

In sum then we consider the following cases:

- The case in which only the seller learns
- The case in which only the buyer learns
- The situation in which the total supply was equal to the optimum demand
- The situation in which the total supply was below the optimum

Finally we looked at the situation where both sides of the market learn.

15.18 Results with a Large Supply

1. *Buyers learn and sellers are passive.*
   Consider the first case in which there is enough supply to meet the optimal demands of the buyers. The buyers learn but the seller has a reservation price fixed at 0. In this situation we consider firstly the case where all the buyers have a reservation price of 2. What one would expect if our system were at all realistic is that the prices would fall to 0 and this is precisely what happens in both the auction and the decentralised mechanisms as can be seen in Fig. 15.18. After 4,000 periods in the bilateral bargaining case and 6,000 in the auction situation. This is true whether the buyers were homogeneous or heterogeneous. They learn to extract the surplus from the passive seller who in the negotiation situation accepts all offers greater than 0 and in the auction case from the auctioneer who does not withdraw the goods at any positive price.
2. *Sellers learn and buyers are passive and homogeneous.*
   Next consider the case in which the seller or auctioneer learns but the buyers are homogeneous and all have a fixed strategy which is to buy at any price below a fixed value, taken here to be 1.5. If our learning process is reasonable the seller

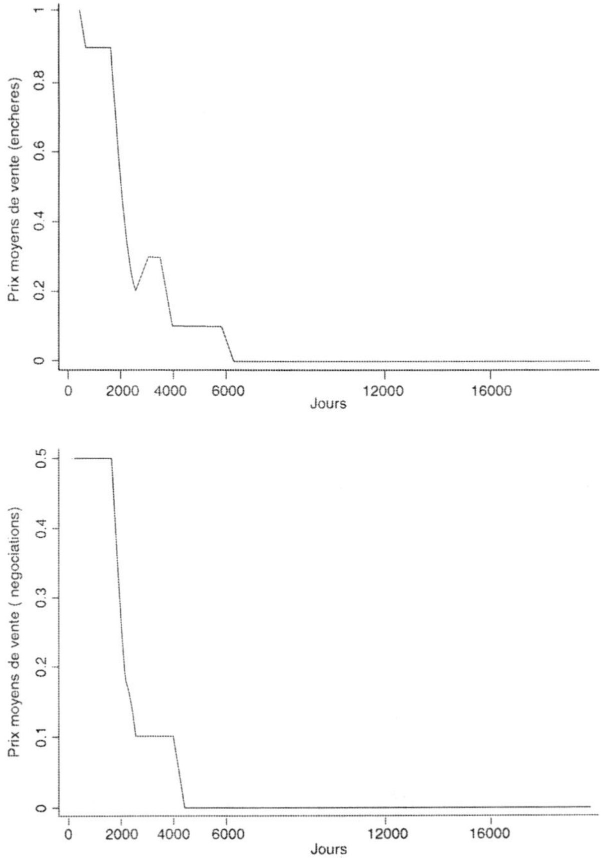

Fig. 15.18 In both the negotiation case (lower panel) and the auction case (upper panel), prices converge to 0. (source Kirman and Moulet (2008))

or auctioneer will learn to extract 1.5 from the buyers. What happens after 4,000 periods is that the opening prices for the auction lower than 1.5 disappear and the minima above 1.5 disappear in the negotiation case. In Fig. 15.19 we can see that in neither the auction nor the bilateral bargaining case do initial prices above 1.5 offered by the seller or proposed as opening price by the auctioneer get used in the long run. The reserve price and withdrawal price also converge to 1.5. The opening prices above 1.5 continue to be used since the auction descends to that value. Again we see that the side which learns manages to exploit its passive opponents.

3. *Sellers learn and buyers are passive and heterogeneous.*
   Now the next and somewhat more interesting case is that in which the buyers are heterogeneous and have different reservation prices. We have three buyers who have values of 1, 2 and 3 respectively. In both the auction and the decentralised case, the seller learns to open at a price above 3 and close at one below 1. In this case all the goods are sold and each buyer pays his reservation price as in Fig. 15.20.

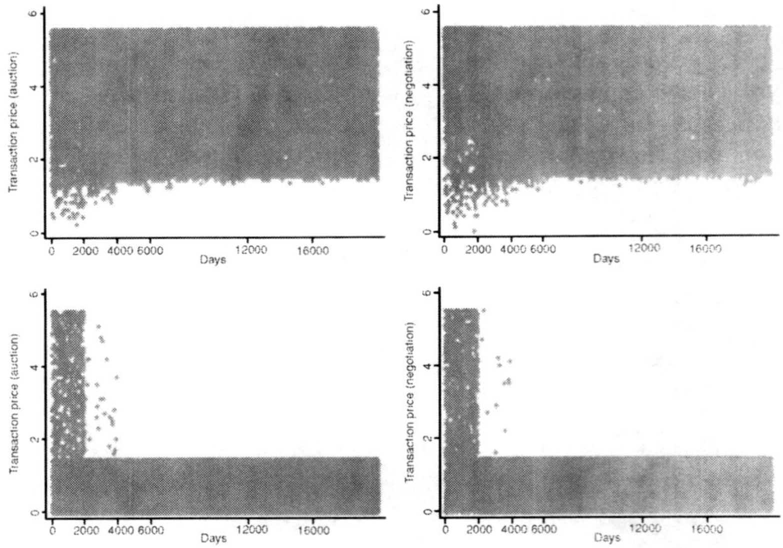

Fig. 15.19 Here we see that in the auction case (*the left panel*) and the bilateral case (*right panel*) prices converge to 1.5. (source Kirman and Moulet (2008))

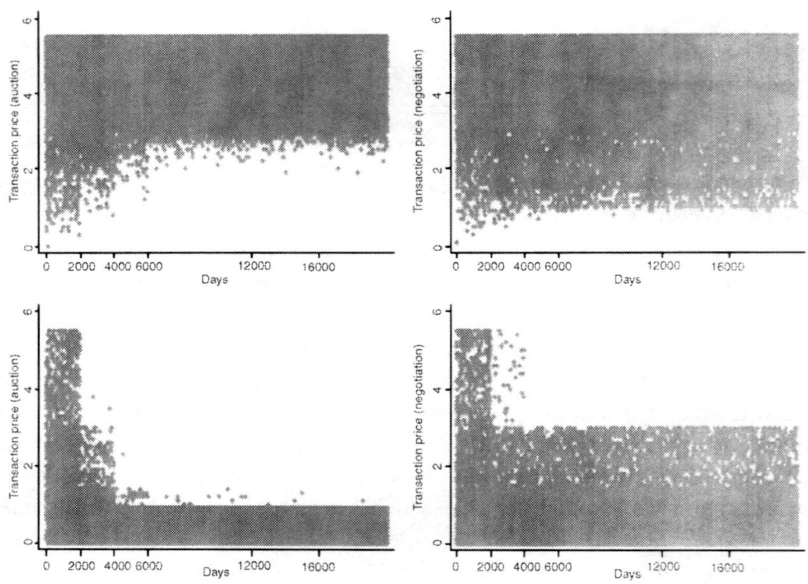

Fig. 15.20 In the two mechanisms, (*auctions left*) and (*negotiations right*) the seller and the auctioneer learn to open at 3 and close at 1 thus extracting all the surplus. (source Kirman and Moulet (2008))

15.19 Results with a Limited Supply

In this case there is competition between the buyers since the total supply is 6. Considering the same cases as before we see that the results are modified.

1. *Buyers learn and are homogeneous and sellers are passive*
   In the decentralised case the prices fall to 0 while at auction converge to 1.15 so the competition prevents the collapse of the market. The goods in the long run are split equally between the buyers.
2. *Buyers learn and are hetergeneous and sellers are passive.*
   Once again in the decentralised case the prices collapse but in the auction situation the prices fall to 1.02. However, now only the two buyers with the higher reservation prices purchase the goods.
3. *Sellers learn and buyers are homogeneous and passive.*
   In this case in both mechanisms the prices settle to 1.5 the reserve price and the seller or auctioneer extracts all the surplus.
4. *Sellers learn and buyers are heterogeneous and passive.*
   Here the two mechanisms give different results. In the case of pairwise bargaining the seller sells at the reservation price of the buyers (3, 1.5, 1) and each buyer buys two units. In the auction situation buyer 1 bids 3 and obtains three units and buyer 2 bids 1.5 and also obtains 3 units. Buyer 3 is excluded from the market and the seller makes a higher profit than in the decentralised case.

15.20 The Market when Both Sides Learn

A number of interesting features appear when we allow both sides of the market to learn.

1. The prices are relatively low given the fact that in the homogeneous case buyers are prepared to bid up to 1.5. Thus the buyers learning process prevents the seller or the auctioneer extracting the surplus.
2. Prices are systematically higher when supply is limited and this effect is particularly apparent in the auction mechanism.
3. In general the auction mechanism generates a higher profit for the seller except in the case of adequate supply in which case the seller earns more in bilateral bargaining.

These results are summarised in Table 15.1.

15.21 Conclusion

The message of this paper is rather simple. Markets are an important feature of all economies. Each market is characterised by an organisation and structure which will have an impact on the outcomes observed. In general it is difficult to capture

Table 15.1 Transaction prices in the various cases. (source Kirman and Moulet (2008)) - Here the prices with the different mechanisms are given after 17,000 rounds during which both sides of the market learn

Supply level	Auction	Bargaining
Homogeneous buyers		
Limited supply (6)	1.22	0.51
Higher supply (9)	0.42	0.46
Heterogeneous buyers		
Limited supply (6)	1.01	0.64
Higher supply (9)	0.38	0.61

all but the simplest feature of such markets in theoretical markets. Even in the case of the simplest markets, those for perishable goods, standard models do not seem to be well adopted to shedding light on the nature of the economic outcomes that one might expect. Curiously enough the particular example which I have used, that of the fish market in Marseilles does exhibit rather a lot of regularity at the aggregate level. Nevertheless this is not due to individuals behaving, in isolation in a regular way as in the standard competitive model. The complicated organisation of this sort of model breaks any simple link between individual and aggregate behaviour. A number of the special features of this market such as the special trading relationships that have developed are difficult to account for in the standard framework. I have suggested first of all, a simple theoretical approach which does capture the formation of such trading relations in a very rudimentary model. To see what happens in a more general case it is necessary to build an artificial market and simulate it. This market exhibits the same features as would have been predicted by the simple theoretical model and which correspond to what is observed on the actual market. To go beyond the problem of trading relations and to examine other aspects of the market such as price dispersion and discrimination I proposed an even simpler approach referred to as "multi-agent" simulation based on very simple rules for learning from past experience and using "classifier systems" to update those rules. This captures the idea that agents learn to use those rules which have been more profitable in the past. This approach is rather successful in reproducing some of the features of the real fish market chosen as an example and shows the link between organisational features and aggregate outcomes.

I then proceeded to examine a market based on a different mechanism, simultaneous Dutch auctions and used the market in Ancona as an example. After an examination of the empirical data and the salient features of this market, I suggested a comparison between the auction market and its bilateral trading counterpart. The simplest versions of the artificial markets used, confirmed what theory would expect but when one allows for both sides of the market to learn some interesting features emerge. At auction, it is not necessarily true that the seller does better in extracting the surplus from heterogeneous buyers than in pairwise trading models, for example.

Artificial markets are particularly useful in studying situations where the interaction and organisation make simple theoretical analysis too difficult. I have tried to illustrate the way in which they can be used so that they cannot be dismissed as ad hoc. Firstly, one can develop a theoretical model in a very restricted case and then simulate an artificial market to see if the conclusions hold up in a more general case. Secondly, one can use a simplified version of the artificial market in which the solution should be obvious to see if it functions correctly before moving on to the more general framework in which the situation is more difficult to predict. This allows us to do more than simply confirm standard theoretical results but also to detect those features which emerge from the additional structure in the artificial markets. Finally these results can be checked with the empirical data.

References

M. Aboulafia. *Making Markets: Opportunism and Restraint on Wall Street*. Harvard University Press, Cambridge, 1997.

M. Aoki. *A New Approach to Macroeconomic Modelling*. Cambridge University Press, Cambridge, 1996.

W.B. Arthur, J.H. Holland, B. LeBaron, R. Palmer, and P. Taylor. Asset pricing under endogenous expectations in an artificial stock market. In W.B. Arthur, S.N. Durlauf, and D.A. Lane, editors, *The Economy as an Evolving Complex System II*, pages 15–44. Addison Wesley, Reading, MA, 1997.

L. Blume. The statistical mechanics of strategic interaction. *Games and Economic Behaviour*, 5: 387–424, 1993.

J. Bulow and P. Klemperer. Auctions versus negotations. *American Economic Review*, 86(1):180–194, 1996.

R.R. Bush and F. Mosteller. *Stochastic Models for Learning*. Wiley, New York, 1955.

C. Camerer and T.-H. Ho. Experience-weighted attraction learning in normal-form games. *Econometrica*, 67:827–873, 1999.

J.M. Epstein. *Generative Social Science: Studies in Agent-Based Computational Modeling*. Princeton University Press, Princeton, 2007.

M. Gallegati, G. Giulioni, A. Kirman, and A. Palestrini. What has that got to do with the price of fish? buyers' behaviour on the ancona fish market. Discussion Paper, Economics Department, University of Ancona, 2007.

A. Greenspan. Corporate governance. Remarks at the 2003 Conference on Bank Structure and Competition, Chicago, Illinois, May 8, 2003.

W. Güth and R. Tietz. Ultimatum bargaining behavior: A survey and comparison of experimental results. *Journal of Economic Psychology*, 11:417–440, 1990.

A. Kirman. Demand theory and general equilibrium: From explanation to introspection, a journey down the wrong road. Annual supplement, History of Political Economy, 2006.

A. Kirman and S. Moulet. Impact de lorganisation du marché: Comparaison de la négociation de gré à gré et les enchères descendantes. Document de travail, GREQAM, Marseille, 2008.

A. Kirman and N. Vriend. Evolving market structure: A model of price dispersion and loyalty. *Journal of Economic Dynamics and Control*, 25:459–502, 2000.

J. Kuhn. *On Todays Business Ethics*. EPIC, Columbia University, New York, 2005.

R. Lucas. Adaptive behaviour and economic theory. *Journal of Business*, 59:401–426, 1988.

A. M. Manelli and D. R. Vincent. Optimal procurement mechanisms. *Econometrica*, 63(3):591–620, 1995.

R. Myerson and M. Satterthwaite. Efficient mechanisms for bilateral trading. *Journal of Economic Theory*, 29:265–281, 1983.

G. Pogrebna. Auctions versus bilateral bargaining: Evidence from a natural experiment. Working paper Columbia University, New York, 2006.

L. Robbins. *An Essay on the Nature and Significance of Economic Science*. Macmillan, London, 1935.

A. E. Roth and I. Erev. Learning from extensive-form games: Experimental data and simple dynamic models in the intermediate term. *Games and Economic Behavior*, 8:164–212, 1995.

H. Simon. The architecture of complexity. *Proceedings of the American Philosophical Society*, 106:467–482, 1962.

G. Weisbuch, A. Kirman, and D. Herreiner. Market organisation and trading relationships. *Economic Journal*, 110:411–436, 2000.

P. Whittle. *Systems in Stochastic Equilibrium*. Wiley, New York, 1986.

L. Xiaohua and R. P. McAfee. The evolutionary stability of auctions over bargaining. *Games and Economic Behavior*, 15(2):228–254, 1996.

Lightning Source UK Ltd.
Milton Keynes UK
UKOW02f1400101214

242899UK00001B/26/P

9 783540 705536